The Quantum Color Code

How Nanocrystals That Glow by Their Size Won a Nobel Prize — and Are Reshaping Medicine, Clean Energy, and the Screens We Live By

Dr. Rosalind K. Thayer

Published by NobleTrex Press

For permissions and other inquiries, write to:
P.O. Box 3132, Framingham, MA 01701, USA

Contents

Introduction

Suppose I handed you three vials filled with what looked like the same substance—clear liquid, nearly invisible particles suspended within it—and told you that one would glow red, one green, and one blue under the right light. You might reasonably assume the colors came from different dyes, different elements, or different impurities. But what if the only meaningful difference was size? Not shape. Not chemical label. Not some artist's hidden pigment. Size. Make the crystal smaller, and it shifts toward blue. Let it grow larger, and it slides toward red.

That sounds less like chemistry than a magician's trick.

And yet this is one of the most important materials discoveries of the last half-century: tiny semiconductor crystals, so small that changing their diameter by a few atoms can change the color they emit. These are quantum dots, and they are among the rare inventions that feel at once deeply abstract and oddly intimate. Their behavior arises from quantum mechanics, the famously counterintuitive physics of the very small. But their effects are not trapped in equations or hidden in laboratory freezers. They are in televisions mounted on living-room walls. They are in biomedical labs where scientists try to make tumors light up more clearly than healthy tissue. They are in experimental solar technologies designed to catch sunlight with unusual efficiency. They are in the color quality of screens we stare at for hours every day, often without the faintest suspicion that the reds and greens are being shaped by crystals only a few nanometers across.

A nanometer is a billionth of a meter. Human hair, that overused benchmark of smallness, is still absurdly large compared with a quantum dot. If a dot were blown up to the size of a basketball, a basketball would swell to something like planetary scale. At that size, matter does not merely become smaller. It becomes different.

That difference is the heart of this story.

We are used to thinking in smooth gradients. A larger pot holds more soup. A shorter ladder reaches less high. A thicker sweater keeps more heat in. Scale usually changes amount, not kind. But the nanoscale is where this intuition starts to fail. Here, the boundaries of an object begin to boss around the electrons inside it. Surfaces become tyrants. Energy, which in bulk materials often behaves as though it comes in near-continuous bands, begins to separate into more discrete possibilities. The old rules don't vanish, exactly; they become fenced in. Shrink a crystal enough, and electrons can no longer ignore the edges. Their allowed states shift. The energy gap changes. And because color is energy made visible, the crystal's glow changes too.

This is why quantum dots seem almost designed to delight the human mind. They are physically real, industrially useful, and conceptually mischievous. They take one of the most notoriously slippery areas of science—quantum mechanics—and pin part of it to a simple, memorable rule: smaller dots, bluer light; larger dots, redder light. It is hard to think of another technology with such a direct line between a childlike question—why is this one blue and that one red?—and a Nobel Prize.

The 2023 Nobel Prize in Chemistry went to Alexei Ekimov, Louis Brus, and Moungi Bawendi for the discovery and synthesis of quantum dots. That phrasing matters. Discovery alone was not enough. Noticing that tiny crystals behaved in size-dependent ways was a scientific revelation. Learning to make them reliably, with control and uniformity, was what transformed revelation into technology. A great deal of modern science advances this way: first an anomaly, then an explanation, then the difficult, unglamorous art of making the thing again and again without ruining it. Quantum dots did not become world-changing because someone proved a point about confined electrons. They became world-changing because chemists learned to cook astonishing precision into beakers of matter.

That achievement is easy to underestimate if your image of chemistry is still the school-lab version: colored liquids, dramatic fumes, and periodic-table trivia. The chemistry of quantum dots is much closer to architecture under pressure. Every nanocrystal is a negotiated structure. Its core composition matters. Its size matters. Its surface matters enormously. The organic molecules attached to that surface matter. The shell added around the core may matter even more. Tiny defects can extinguish light. A good coating can rescue it. The same particle that looks brilliant in a pristine flask may slowly dim, clump, poison performance, or fall apart when exposed to oxygen, moisture, heat, electric fields, or the biochemical chaos of the human body. To make a quantum dot useful, you must control not only what it is, but what it touches, how long it lasts, and what it does when the world refuses to behave like a textbook.

That is one reason this subject matters so much to everyday life. We live in an age shaped not just by information, but by materials subtle enough to control the flow of light itself. The quality of a screen is no longer a trivial consumer concern; it influences how billions of people work, communicate, learn, shop, watch, diagnose, and create. Better displays are not only prettier. They can be more energy-efficient. They can render color more faithfully. They can waste less light in the process of generating an image. A "better red" sounds like marketing language until you realize how hard it is to produce a red that is both pure and efficient, and how many engineering compromises sit behind every saturated pixel.

Medicine, too, has become increasingly visual. Modern biology often begins by attaching light to molecules and asking where they go. If you can label different structures with different colors, you can inspect living systems with far greater nuance. Quantum dots are attractive here because they can be bright, stable, and sharply colored. Instead of bleaching out quickly like some conventional dyes, they can keep shining. Instead of producing broad, overlapping colors that are hard to distinguish, they can emit comparatively narrow spectral signatures. That means many biological targets can, in principle, be tracked at once. A tissue sample can become a map. A tumor margin can become more visible. A diagnostic image can carry more layered information. But biology is not impressed by optical elegance alone. Anything introduced into the body must also survive circulation, find the right target, avoid the wrong one, and ideally leave when its job is done. So the very same features that make quantum dots exciting also force difficult questions about toxicity, clearance, and trust.

Then there is energy. Sunlight arrives as a broad spectrum; conventional materials harvest only selected parts of it efficiently. Quantum dots offer a tantalizing possibility: absorbers and emitters whose behavior can be tuned by design, as if color were not merely a consequence of a material, but one of its adjustable settings. That could matter in solar cells, where matching a material's electronic structure to incoming light is central to efficiency. It could matter in luminescent solar concentrators, where windows or panels absorb light and re-emit it in directions useful for power generation. Once again the promise is seductive and the obstacles are real. Capturing light is one thing. Getting charges out efficiently, keeping materials stable outdoors, and manufacturing them cheaply at scale is another.

That tension—between dazzling physical simplicity and stubborn practical complexity—is what makes quantum dots such a rich subject. The central principle can fit on a postcard. The consequences fill industries.

And the story is not merely about one class of tiny crystals. It is about a broader shift in how humans make materials. For most of history, we accepted the properties nature gave us. Wood burned, copper conducted, glass transmitted, iron rusted, and pigments reflected whatever colors their chemistry allowed. Modern materials science changed the terms of that relationship. It taught us to ask not just, "What is this made of?" but "How is it structured, and at what scale?" Quantum dots are among the purest expressions of that new mindset. They reveal that geometry itself can become a control knob for physics. Change the dimensions of a crystal, and you rewrite its behavior.

That idea borders on philosophical. It says that matter is not only composition but confinement. Not only ingredients but arrangement. Not only substance but size.

It also says something humbling about the world. The vivid image on an expensive display, the glowing marker in a pathology lab, the prototype coating for a solar device—these are not triumphs of brute force. They are triumphs of restraint. Quantum dots work because electrons are denied their usual freedom. They are hemmed in, quantized, and made to reveal their options one by one. There is poetry in that, if you like your poetry with centrifuges and absorption spectra.

Over the pages ahead, we will keep returning to a few deceptively simple questions. Why should size determine color at all? What happens inside a semiconductor when it becomes so small that "bulk" no longer means much? How do chemists manufacture millions upon millions of nearly identical nanocrystals when tiny variations would blur the very color purity they seek? Why do some dots shine with extraordinary efficiency while others waste their energy in silence? What makes a quantum dot ideal for a television and troublesome for a bloodstream? Why did it take decades for an elegant laboratory phenomenon to become a practical commercial platform? And why, even after the Nobel recognition, does the field still feel unfinished?

The answers connect disciplines that are too often taught as separate worlds. Physics explains why confinement creates discrete energy behavior. Chemistry explains how to make and stabilize that behavior. Engineering decides whether the result survives heat, time, oxygen, manufacturing tolerances, and customers. Biology and medicine ask whether the same glowing particles can communicate with living tissue without doing harm. Policy and environmental science ask what happens when brilliance has a toxic ingredient list. Industry asks a rude but necessary question that science alone cannot answer: can you make it by the ton, not just by the paper?

The most exciting science often begins when an object refuses to fit into the categories we inherited. A quantum dot is not quite a molecule and not quite a chunk of solid. It is too large to behave exactly like one, too small to behave comfortably like the other. It occupies a middle kingdom where collective material properties and discrete quantum rules overlap and argue. That ambiguity is not a nuisance. It is the source of the magic.

And magic, properly understood, is usually just precision we haven't learned to appreciate yet.

So look for quantum dots not only in the laboratory but in the habits of modern life. In the blue LED backlight converted into richer reds and greens on a screen. In the fluorescence microscope sorting cellular components by color. In a solar concept trying to shepherd photons more cleverly. In regulations that quietly push researchers away from cadmium and toward safer alternatives. In factories where batch-to-batch uniformity becomes as important as the original insight. In the weirdly moving fact that a few nanometers can determine whether something shines ruby, emerald, or sapphire.

The story begins with color, because color is where quantum dots first announce themselves. But the deeper subject is control: control of electrons, of surfaces, of defects, of synthesis, of interfaces, of risk, of scale. A quantum dot glows by its size, yes. Yet what made these nanocrystals revolutionary was the realization that size could become a language—a way of instructing matter to do something exquisitely specific.

Once that language existed, people found themselves saying all kinds of ambitious things with it.

That is where the surprise really lies. Not that tiny crystals can glow different colors, though that remains delightful. It is that this odd nanoscale rule has become one of the clearest demonstrations that the future may depend less on discovering entirely new substances than on learning to shape old ones with unprecedented finesse. A semiconductor is familiar. A crystal is familiar. Light is familiar. But shrink the crystal until quantum mechanics takes over, and familiarity turns into a tool kit.

A pocket rainbow in your living room. A fluorescent scout in a tumor sample. A tunable absorber in a solar device. A Nobel Prize awarded for proving that when matter becomes small enough, color itself can be engineered.

That is not a side note in modern science. It is one of its signature ideas.

Chapter 1

A Crystal That Picks Its Own Color

A modern screen can look almost impossibly saturated, yet its most remarkable trick begins in particles so small they blur the boundary between chemistry and physics. Follow that trail inward and a strange idea appears: make a crystal smaller, and it does not merely become less of itself—it becomes a different color, a new kind of object, and a clue to how the nanoscale rewrites the rules.

1.1 The Pocket Rainbow in Your Living Room

A modern screen can feel like a small miracle performed at arm's length. A red sports car glows with a lacquered intensity that older televisions never quite managed. A rainforest documentary suddenly contains greens you didn't know your living room could hold: not just "green," but the sharp, almost electric green of new leaves backlit by sun. Even the little icons on a phone can look as if they were painted with freshly opened pigments rather than lit from behind.

It is tempting to treat this as a style choice, the way fashion drifts toward brighter or duller colors. Yet the real story is closer to the history of electric light than the history of art. Those reds and greens are hard-won feats of control over matter, and they depend on a surprisingly picky demand: if you want a display to make a truly vivid red, it helps if your red is *only* red.

That sounds obvious until you remember what "red light" usually means in technology. A conventional "red" source often sprays out a whole neighborhood of wavelengths: mostly red, but also some orange, maybe a tail toward yellow, perhaps even a bit of deep red edging into the near-infrared that your eyes can't see but your device still pays for in wasted energy. Engineers describe this spread with a spectrum, and what they long for is a narrow peak, a tight little hill centered exactly where it should be, with steep sides that don't spill into neighboring colors.

Why does that matter? Because displays, at heart, are juggling acts with only a few balls.

Nearly all screens create color by mixing three primaries: red, green, and blue. A pixel is really a tiny stage where these three lights perform in different intensities. When the red, green, and blue are well chosen, their mixtures can approximate a wide range of colors: the subtle tans of skin, the slate blues of evening shadows, the murky green-brown of a river after rain.

But the range you can reach depends on how *pure* the primaries are. If your "red" includes a chunk of orange and yellow, then when you try to mix it with green to make certain shades, the mixture gets muddied. In the geometry used by display designers, the most saturated primaries define a larger "gamut"—a bigger patch of the color map that the screen can cover. The everyday translation is simple: purer primaries give you more colors that look convincing, and the most noticeable improvements often appear at the extremes. People say, "The reds pop," not because they became art critics overnight, but because the old compromises were most visible there.

The path to these purer primaries runs straight through the awkward fact that screens are not just making color; they are managing light like a budget.

Take a common kind of television: the LCD. An LCD does not produce light by itself. It is a clever shutter array. Behind it sits a backlight, and the job of the LCD is to let through more or less light at each pixel while color filters select red, green, or blue portions. If you start with a "white" backlight, you might assume you're in good shape: white contains all colors. Yet most white backlights are not truly broad, smooth rainbows. For many years the workhorse was a blue LED plus a yellowish phosphor. The blue LED provides a strong blue peak; the phosphor converts some of that blue into a broad swath of longer wavelengths that your eyes interpret as yellowish. Together, the blue and the broad yellow can look white.

The problem comes when you pass that light through red and green filters. Filters are picky: they keep a narrow slice and throw away the rest. If your backlight's spectrum is messy, the filter discards a lot of it as heat. To get a bright image, you crank up the backlight, spending more power and creating more heat for the same number of useful photons. Energy efficiency and color quality end up tangled together.

This is where quantum dots slipped into living rooms, not with fanfare about quantum theory, but as a practical hack to make better red and green light from a blue LED.

The idea is elegantly simple in outline. Keep the blue LED backlight, because blue LEDs are efficient and robust. Then place in front of it a thin layer containing quantum dots engineered to absorb some of the blue and re-emit it as very narrow red and green. What comes out is still a mixture that can serve an LCD: blue that passed through, plus new red and green created by the dots. But now the red and green are sharply defined peaks rather than wide, sloppy bands. The color filters have an easier job: they discard less, so you waste less. The screen can hit deeper reds and cleaner greens without brute-force brightness.

A thin layer that does this job is often called a quantum-dot enhancement film. The name sounds like marketing, but it describes a real component: a film, typically only a fraction of a millimeter thick, embedded with nanocrystals that behave as efficient color converters.

A vivid screen is the consumer-facing result of two underlying virtues that engineers obsess over: *narrow emission* and *high efficiency*.

Narrow emission is about where the light goes in color space. High efficiency is about whether the light comes out at all.

If you shine blue light on a quantum dot and it responds by emitting a red photon, you have converted one color into another. But the conversion is not guaranteed. The dot might absorb a photon and then lose the energy without emitting light, perhaps as vibration and heat. The fraction of absorbed energy re-emitted as light is called the photoluminescence quantum yield. A dot with a high quantum yield is bright; a dot with a low one is a disappointment, no matter how perfect its color might have been on paper.

This pairing of spectral purity and efficiency helps explain why "better red" matters so much. Our eyes are not equally sensitive to all colors. They are less sensitive in the deep red. If your red primary is slightly orange-ish or broadened, you may need to turn it up to get the same perceived saturation, and that costs power. A narrow, properly placed red can achieve a strong perceptual effect with fewer wasted photons. In a battery-powered phone, that can matter. In a wall-sized television that runs for hours each evening, it matters in a different way.

The story has a historical echo. When cities first electrified, light quality competed with efficiency. Early lamps could be harsh and bluish or warm and dim. Engineers learned that the spectrum of light is not a luxury; it is a design parameter with economic and social consequences. Screens have been going through a similar maturation, except the "lamp" is now a layered optical system, and the spectrum is being sculpted by particles thousands of times smaller than a grain of dust.

Once you start looking, the same sculpting shows up in places far from the couch.

Walk into a biology lab and you may see a rack of tiny tubes that look as if they contain jewel-toned inks. Under ordinary room light, they might be faintly colored or even nearly clear. Under ultraviolet or blue illumination, they glow with startling brightness: green, yellow, orange, red. To a biologist, this isn't decoration; it is information. Each color can be attached to a different biomolecule, so that several targets can be tracked at once in the same sample. Color becomes a multiplexed language: one hue for a receptor on a cell surface, another for a protein inside the nucleus, another for a drug molecule moving through tissue.

Traditional fluorescent dyes made this sort of labeling possible long before quantum dots arrived, and a good dye is still a beautiful piece of chemistry. But dyes have limitations that become obvious when you try to do several colors at once, or when you try to image for a long time. Many dyes photobleach: they permanently stop fluorescing under illumination, like a T-shirt left too long in the sun. Their emission spectra can also be broad, which makes it harder to separate multiple labels cleanly. Overlapping colors are not just aesthetically unpleasant in a scientific image; they can scramble a measurement.

Quantum dots offered a different set of trade-offs. Because their emission peaks can be narrow and bright, they can make it easier to separate channels: green means green, not a green with a lot of yellow tail that bleeds into the next filter band. And because many quantum dots can be excited with the same light source while emitting in different colors, a single excitation lamp can drive a multicolor experiment. A researcher can illuminate at one wavelength and read out several distinct emission colors, each tied to a different target. That is the same basic advantage the LCD backlight exploits: one blue source, multiple precisely placed outputs.

A useful anecdote from the early days of nanocrystal science captures how strange this "color as design" idea once felt. In the late twentieth century, scientists including Alexei Ekimov and Louis Brus noticed that tiny semiconductor crystals behaved differently from bulk material, showing shifts in optical features as their sizes changed. This was not merely a curiosity about small particles; it suggested a new kind of manufacturing knob. Instead of choosing a pigment or a phosphor by its chemical identity alone, you could choose it by its *size*. The same substance could be tuned across colors like a set of violin strings tightened to different pitches.

For a long time, this remained the sort of result that delights physicists and frustrates engineers. A laboratory vial that glows beautifully is not the same as a material you can put into a factory process. Displays are especially unforgiving. If you coat a film with emitters whose colors vary even slightly from particle to particle, you get a broad spectrum, and broad spectra mean less saturated colors. If the material degrades under heat or oxygen, the picture changes over time. If it is too sensitive to moisture, your manufacturing yield collapses. Even if the dots are brilliant in solution, they can clump or lose brightness when embedded in a solid film.

The transformation from "pretty vial" to "shipping product" depended on chemists learning to make quantum dots with remarkably tight control over size and surface chemistry. One landmark on that road was the development of reliable colloidal synthesis routes that could produce nearly monodisperse nanocrystals—meaning the particles are nearly all the same size, rather than a wide mix. In ordinary life, we rarely care if grains of sand differ in diameter by a few percent. In quantum-dot optics, that few percent can be the difference between a sharp red and a smeared, orangey mess. Uniformity is not perfectionism; it is how you prevent the spectrum from blurring.

Once you appreciate that, a screen looks less like a slab of glass and more like a treaty between several competing demands. You want your red and green peaks narrow and in exactly the right places. You want high quantum yield, so blue photons aren't lost as heat. You want the converter layer to be stable under years of illumination and the warmth of a running device. You want manufacturing consistency across millions of units. And because these materials sit in consumer products, you also want them to be acceptable under environmental and safety regulations, which have shaped choices about which elements can be used and how they must be contained.

All of that is hidden behind the ease of a remote control.

There is a subtle psychological twist here: better color is not only about spectacle. It can be about *honesty*. When a display can't reach a saturated green, it substitutes the closest color it can make, and the substitution is not neutral. It changes how we perceive a landscape photo, a medical image, a piece of clothing on an online store, even a national flag on a news broadcast. Color is part of meaning. A screen that can produce a truer red can make a tomato look ripe rather than dull; a screen that can separate shades of green can make a heat map easier to read. These are small effects, but small effects multiplied by billions of screens become a background condition of modern life.

The most revealing way to think about quantum dots in this context is not as tiny light bulbs, but as tiny contracts: for a given input of blue light, they promise to return light in a narrowly defined color, and to do it efficiently. That contract is written in the physics of confined electrons and in the chemistry of surfaces that can either encourage light emission or sabotage it. When the contract is honored, the result is a pocket rainbow: a controlled set of spectral lines bright enough for a living room, precise enough for a laboratory assay, and reliable enough to manufacture at industrial scale.

A screen that makes you pause—that moment when a red looks impossibly red—is a reminder that "color" in technology is not paint. It is a carefully engineered distribution of photon energies, negotiated between quantum behavior and practical constraints, and delivered to your eyes through a stack of materials that has learned, at last, to waste less of its light.

1.2 When Small Stops Behaving Like Smaller

Miniaturization usually sounds like an accountant's kind of progress: the same thing, made smaller, cheaper, faster. A smaller transistor is still a transistor; a smaller camera is still a camera. That intuition works for a long time because most objects we shrink remain governed by the same rules. A toy piano has fewer keys than a concert grand, but the notes it produces still follow the familiar logic of vibrating strings and resonant wood.

A nanocrystal breaks that intuition. Make a semiconductor crystal small enough and it stops behaving like a shaved-down chunk of ordinary solid. The change is not cosmetic. It is a shift in what the electrons are *allowed* to do.

Start with a picture of a bulk semiconductor: a piece of silicon, or cadmium selenide, or gallium arsenide, large enough that its interior atoms have neighbors on all sides, repeating in an orderly lattice. In that bulk material, an electron is not tied to one particular atom. It can be shared across many atoms, described by a wave spread throughout the crystal. Physicists express this by saying electrons occupy *bands* of allowed energies rather than isolated atomic levels. In everyday language, a bulk crystal offers electrons a vast dance floor with many ways to move.

Now imagine carving that crystal down to a piece only a few nanometers across. At that point, two things happen at once. First, the crystal has hardly any "interior" left. Second, the electron's wave-like nature, so easy to ignore when the floor is huge, becomes impossible to ignore when the room is cramped.

The surface problem comes first, because it is geometrical and almost unfairly simple. If you build a cube out of atoms, the number of atoms on the surface grows like the area, while the number inside grows like the volume. Shrink the cube and the surface becomes a larger and larger fraction of the whole. In a macroscopic crystal, the surface is a thin skin on a big body; in a nanocrystal, the surface is a major organ.

That matters because surfaces are messy in a way interiors are not. An atom in the interior has its full set of neighbors, locked into a repeating pattern. An atom on the surface has missing neighbors and "dangling" bonds—unfinished chemical handshakes. Those dangling bonds can introduce electronic states that are not part of the clean bulk bands. They can trap electrons or holes (the absence of an electron that behaves like a positive charge carrier) and provide routes for energy to leak away without light. Surface atoms also see the outside world: solvents, oxygen, water, ligands (molecules bound to the surface), neighboring particles in a film. So when nanocrystals behave differently, part of the reason is simply that so much of the material is no longer protected by the calm regularity of the interior.

Even if the surface were magically “perfect,” smallness would still change the rules because electrons are waves as well as particles. A wave does not fit into a confined space in arbitrary ways. If you pluck a guitar string, it cannot vibrate with just any pattern; it must match the boundary conditions at the fixed ends. Only certain standing waves are allowed, each with its own frequency. A drumhead, a flute, and a violin all share this same logic: confinement forces vibration into discrete modes.

Electrons in a crystal are not sound waves, but the same kind of constraint applies. An electron wave spread through a solid must satisfy boundaries too. In a large crystal, the boundaries are so far away that they barely matter, and the allowed energies crowd together so densely that they look continuous. In a tiny crystal, the boundaries are close. The allowed patterns become more limited and the energies spread apart. Instead of a near-continuum of possibilities, you begin to get a ladder with rungs you can, in principle, count.

This is the core surprise behind quantum dots’ size-tunable color: not a new chemical composition, but a new set of allowed electronic motions created by confinement.

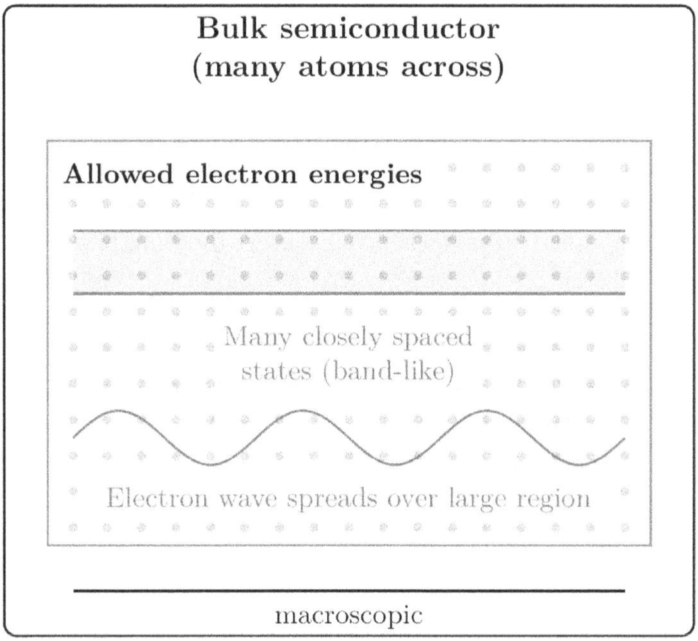
Bulk semiconductor
(many atoms across)
Allowed electron energies
Many closely spaced
states (band-like)
Electron wave spreads over large region
macroscopic

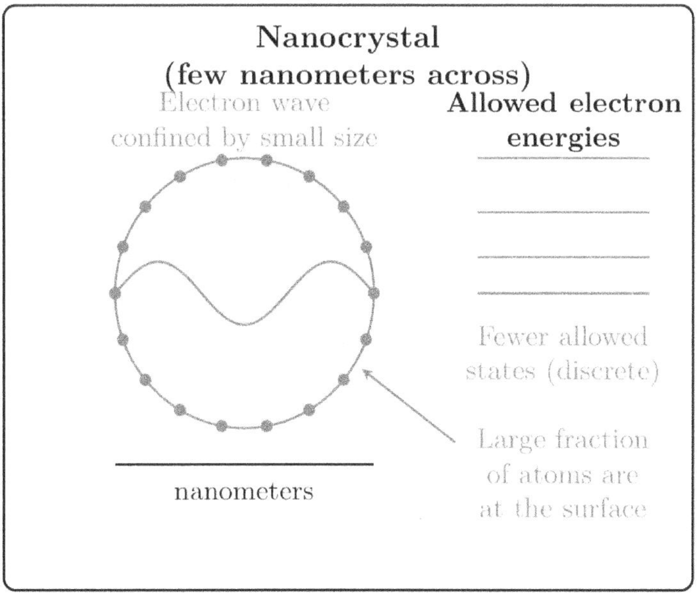
Nanocrystal
(few nanometers across)
Electron wave
confined by small size
Allowed electron
energies
Fewer allowed
states (discrete)
Large fraction
of atoms are
at the surface
nanometers

The confinement idea can be made intuitive without equations by thinking in terms of "wiggle room." A wave that has lots of space can vary gently. A wave forced into a small region has to bend more sharply to satisfy the boundary conditions, and sharper bends correspond to higher kinetic energy. That is the reason confinement raises energies: not because you added energy, but because the allowed states are pushed upward when you squeeze the wave into a smaller space.

There is a famous, almost pedagogical example in early quantum mechanics called the "particle in a box," where a quantum particle trapped between walls can only occupy certain energy levels. Real nanocrystals are not literally hard-walled boxes. Their boundaries are softer, and the material outside the dot matters. But the basic lesson survives: making the dot smaller increases the spacing between allowed energy states.

In a semiconductor, light absorption and emission often involve an electron being promoted from a lower-energy band (the valence band) to a higher-energy band (the conduction band), leaving behind a hole. The electron and hole can move under the influence of the crystal's periodic lattice and can attract each other, forming a pair called an exciton. In a bulk semiconductor, that exciton has a characteristic size: it is spread over a certain distance, set by the material's properties. If the crystal is large compared to that exciton size, the pair behaves much as it would in the bulk. If the crystal becomes comparable or smaller, the exciton no longer fits comfortably. It is squeezed, and the energies shift.

This is the moment where "smaller" stops being merely "less." The material has not changed its chemical identity, but the electronic structure that determines optical properties has changed. The dot has become, in a meaningful sense, a different optical object.

The history of how this realization emerged contains a nice reminder that science often advances when someone is paying attention to an anomaly rather than smoothing it away. In 1981, Alexei Ekimov and colleagues reported that semiconductor microcrystals grown in glass showed a systematic shift of absorption features to shorter wavelengths as the crystals got smaller. Shorter wavelength means higher photon energy: a move toward the blue. It was a clue that the energy levels were not fixed by chemistry alone. They were moving with size.

A few years later, Louis Brus (1984) provided a theoretical description of how the lowest excited state in small semiconductor crystallites should depend on size. The key idea was that confinement contributes a term that rises strongly as the radius shrinks, while the Coulomb attraction between electron and hole pulls the other way, partly offsetting that rise. Even if one never writes down the famous simplified "Brus equation," the competing physical influences are conceptually valuable: squeezing the wavefunctions tends to increase energy, but bringing charges closer also changes their interaction. The observed color, then, is not a simple ruler attached to diameter alone; it is a negotiation among confinement, attraction, and the particular material's internal constants.

That nuance is not academic fussiness. It helps explain why different semiconductor materials span different spectral ranges, and why dots of the same size but different composition can emit different colors. It also clarifies why the same dot can shift slightly depending on what surrounds it: the boundary is not just a geometric edge, but an interface between materials with different abilities to host an electron wave. In practice, the dot and its environment form a coupled system.

Surface chemistry, once again, becomes central. The ligands that bind to a nanocrystal are not mere packaging peanuts. They can change how well the dot is electronically isolated from its surroundings; they can heal dangling bonds; they can introduce or remove electronic trap states. If a trap state sits energetically in the wrong place, it can intercept an excited electron or hole before light is emitted. The dot absorbs light, but then the energy drains away as heat. This is one reason the same nominal "dot" can be dazzling in one synthesis and disappointing in another: the internal core may be similar, but the surface may be electronically night-and-day different.

The dominance of surfaces at the nanoscale has an everyday analogy that feels almost too simple: peeling an apple. For a large apple, the peel is a thin fraction of the total. For an apple the size of a pea, the peel would be a large fraction of what you eat, and its taste would matter much more. Nanocrystals are like that pea-sized apple, except the "peel" is not a culinary detail but the frontier where electronic states can be created, destroyed, and modified.

There is another way to feel the shift in regime: imagine trying to describe a city by talking only about its downtown. In a huge metropolis, the downtown may dominate the economy and culture. In a small town, the "edge" is everywhere. The boundary conditions define the character. A nanocrystal is a small town for electrons: the edge is always near, the outside world always influential.

This helps make sense of why quantum dots can look so crisp in color when they are well made, and so disappointing when they are not. The physics promises discrete, size-dependent electronic states; the chemistry decides whether those states are visible as light or hidden behind non-radiative losses. A dot is simultaneously a tiny crystal and a tiny interface problem.

The deeper point is that the nanoscale is not merely a smaller stage for the same actors. It is a stage where the script changes. Electrons and holes cease to behave as if they are moving through an effectively infinite solid and start behaving as if their possible motions are rationed. Surfaces cease to be negligible and start to dominate. At that point, changing a dimension by a nanometer is not a manufacturing tolerance; it is a way of rewriting the allowed quantum states, and therefore rewriting the color the crystal can produce.

1.3 Color as a Ruler

A jeweller learns to judge a diamond by how it handles light: the flashes, the fire, the way a stone turns a white lamp into tiny rainbows. With quantum dots, light is not just a way to admire the object. Light is the measuring instrument.

That sounds like metaphor, but in practice it is literal. In a laboratory, the first "measurement" of a newly made batch of dots is often not a ruler, not a microscope, not an image at all. It is a spectrum: a record of which colours the dots absorb and which they emit. From that spectrum, a skilled chemist can often tell whether the synthesis has gone well, whether the particle sizes are tightly clustered, and roughly how large the dots are.

The reason this works is the same reason the dots are useful in displays and imaging: their optical energies are linked to their size. As the dots shrink, confinement pushes the relevant electronic energies upward; as the dots grow, those energies slide downward. The result is a reliable direction of change: smaller dots correspond to higher-energy (bluer) light, larger dots to lower-energy (redder) light. What matters for measurement is not only the direction but the repeatability. If the relationship between size and colour were fickle, colour would be a party trick. Because it is systematic, colour becomes a ruler that can read nanometres.

Two related experiments sit at the heart of this "ruler."

One is absorption. Shine a broad range of light through a solution of dots and some wavelengths will be eaten more than others. The absorbed wavelengths promote electrons to higher-energy states. For many semiconductor dots, the onset of strong absorption and the position of the first distinct absorption feature (often called the band-edge absorption) shift with particle size. Ekimov's early glass-matrix studies (1981) already contained this clue: as microcrystals became smaller, their absorption features shifted to shorter wavelengths, meaning higher energies.

The second is emission, usually measured as photoluminescence. You excite the dots with a lamp or laser—often in the ultraviolet or blue—and then collect the light they give back. Emission is not the same as absorption. Absorption tells you what energies the dot can take in; emission tells you what energy it prefers to give back after it has had time to relax internally. The emitted colour is often narrow and vivid, which is part of why quantum dots are prized. For measurement, the key point is that the peak emission wavelength also shifts with size in the same systematic direction.

The absorption spectrum and emission spectrum are like two kinds of fingerprints. They are related but not identical, and reading them well requires knowing what each one is saying.

Absorption is often the more direct window onto the dot's allowed excited states. It can show multiple features—little shoulders and peaks—that correspond to transitions between quantized states. Emission, by contrast, usually comes from the lowest available excited state after the dot has "cooled" internally. A dot can absorb a higher-energy photon, shuffle that energy among its electrons and vibrations, and then emit from the lowest excited state available. This is why emission typically appears at a slightly longer wavelength than the absorption edge; some energy is lost to internal relaxation before a photon is emitted. The gap between absorption and emission peaks is called the Stokes shift, and while it is not the star of the story here, it is a useful reminder that emission is a *process*, not a static label.

What makes colour a ruler rather than a mere label is that it can be calibrated. If you can connect a particular emission peak to a particular mean size for a given material system, then a spectrum becomes an indirect size measurement. "Indirect" is not a weakness; it is often a strength. Electron microscopes can measure size directly, but they are slow, expensive, and not always representative—especially if particles clump on a grid or change during preparation. Optical spectra can be taken in minutes, on a tiny sample, in the same solution the dots will later be processed from. For manufacturing, that speed and convenience are decisive.

The catch is that rulers are only useful when the marks are sharp.

A batch of dots is never perfectly uniform. There is always some distribution of sizes. If that distribution is wide, the emitted colours from individual dots spread out and pile on top of one another, producing a broad emission peak. A broad peak is not only less desirable for a display; it is a blurry measurement tool. It is like trying to weigh flour with a scale that jitters by a few grams.

Monodispersity—the degree to which particles are the same size—is therefore central to using colour as measurement. High monodispersity gives narrow absorption features and narrow emission peaks. Low monodispersity smears everything out. Ekimov's early work in glass already encountered this practical complication: dots formed with a range of sizes, and that dispersion made interpretation harder. The physics was present, but it arrived wrapped in variability.

Chemistry eventually supplied the missing piece: synthetic methods that produced dots with much tighter size distributions. A particularly influential step was the development of hot-injection colloidal synthesis methods that separated nucleation (the birth of new particles) from growth (the enlargement of existing ones). In the landmark work by Murray, Norris, and Bawendi (1993), a rapid injection created a short burst of nucleation; afterward, growth proceeded more uniformly on the already-formed seeds. When nucleation and growth overlap for a long time, the batch ends up as a mix of newborn tiny dots and older larger dots. When nucleation is brief and growth is controlled, the batch can become nearly monodisperse. Their work also emphasized size-selective precipitation—using solvents to coax out particles by size—further tightening distributions.

This shift turned optical spectra into something like a well-marked ruler. Narrow absorption features made it possible to say, with confidence, that a given peak corresponded to a particular characteristic size and to track growth in real time. Chemists could take small aliquots during a synthesis, measure the emission peak, and decide whether to let the reaction run longer or quench it. The dot's colour became feedback, and feedback became control.

That feedback loop is one of the quiet revolutions behind modern quantum-dot technology. The public sees a television advertisement boasting richer reds. A chemist sees something else: a clean spectrum in the fume hood at 2 a.m., signalling that the reaction has landed on the right size and stayed there with minimal spread.

The basic mapping between size and colour is easy to state, and difficult to overstate in its consequences: within a given semiconductor material system, smaller dots emit bluer light and larger dots emit redder light. The emitted photon energy tracks an effective band gap that increases as confinement grows stronger. The same principle shifts absorption as well: the absorption onset moves to higher energies (shorter wavelengths) as dots get smaller. Brus's theoretical work (1984) gave a framework for this size dependence by balancing a confinement-driven rise in energy with Coulomb attraction between electron and hole that partially offsets it. In everyday terms, squeezing the wavefunctions costs energy, but bringing opposite charges closer also changes their interaction.

A plot like this, in a lab notebook, is more than a tidy summary. It is also a reminder that "size" is not a single number. A real sample has a mean size and a width to its distribution. The width shows up optically as linewidth: the spread of wavelengths around the peak. Narrow linewidth signals high monodispersity; broad linewidth suggests a mixture of sizes, or additional complications such as emission from surface-related states. The peak position is the ruler mark; the width is the uncertainty.

The difference is immediately felt in practical applications. For a display backlight, a broad emission peak means less saturated colour and more overlap between channels. The red leaks toward orange; the green leaks toward yellow. The colour filters in the panel must work harder, throwing away more light, and the achievable gamut shrinks. For a biological label, a broad emission peak makes it harder to separate multiple fluorophores cleanly; two labels whose emissions overlap can "bleed" into each other's detection channels, a problem that becomes severe when signals are weak.

This is why the chemistry that tightens size distributions is not merely a manufacturing convenience. It is part of the optical principle itself. A ruler with thick, fuzzy tick marks is not much of a ruler.

There is a second subtlety: the same colour does not always mean the same size across different materials, and even within one material system, the surrounding environment can nudge the spectrum. A dot's surface ligands, the solvent, and whether the dot is isolated in solution or packed into a solid film can shift energies slightly. The "ruler" is therefore best understood as a calibrated tool used under defined conditions. In a careful laboratory, calibration is a living practice: compare optical measurements with direct size measurements from microscopy, update the mapping, check that processing into films does not introduce unexpected shifts.

Absorption can sometimes be the more robust indicator for size during synthesis because it is less sensitive to certain non-radiative losses. A dot can absorb light even if it emits poorly; emission, by contrast, depends on whether the excited state ends up producing a photon rather than being quenched by surface traps or other defects. This is why chemists often glance first at an absorption trace to judge growth and then look at photoluminescence to judge surface quality. The two measurements together help distinguish "the size is wrong" from "the surface is wrong," which can look deceptively similar if one only stares at a dim vial.

The ruler metaphor also clarifies why the earliest observations were tantalizing but incomplete. Seeing absorption shift with size, as in Ekimov's 1981 work, is like discovering that a thermometer responds to temperature but comes with an irregular scale and a sticky mercury column. The principle is there, but it cannot yet be used with confidence. Brus's 1984 theory made the scale intelligible, even if only approximately. The synthetic advances of the early 1990s made the column smooth and the scale readable. Only then could the colour-size mapping become a dependable tool for both science and technology.

It is hard to overemphasize how strange this is by the standards of ordinary materials. A bar of copper does not change colour because you filed it thinner. A bulk semiconductor has a fixed band gap set by its composition and crystal structure. But a semiconductor nanocrystal lets you tune an optical property by adjusting a dimension measured in nanometres. That turns manufacturing into a kind of spectroscopy and spectroscopy into a kind of manufacturing. You "dial in" a target emission colour by controlling nucleation, growth, and purification, and then you verify that you hit the target by measuring the emitted light.

Once that habit of thought takes hold, it becomes natural to see a glowing sample as a report rather than a decoration. The colour is the system telling you what state it is in. The sharpness of the colour is the system telling you how uniform it is. And the brightness is the system telling you whether the surface is cooperating. A small vial under a blue lamp can therefore carry a surprising amount of information: a nanometre-scale measurement encoded in visible light, legible to the human eye, and precise enough to guide the next step of chemistry.

1.4 Three Battles Every Good Dot Must Win

A nanocrystal can have the right colour and still be useless.

That blunt fact tends to surprise people because colour feels like the whole point. If the size sets the hue, and chemistry can tune the size, what else is there? In practice, the size—colour relationship is the beginning of the job, not the end. Real devices and real experiments demand something stricter than "red-ish" or "green-ish." They demand sameness, brightness, and reliability—not once, but in every batch, in every factory run, after months on a shelf, after years under light.

Those demands are not a matter of consumer fussiness. They are consequences of the physics that makes size-tunable emission possible. When the electronic states are sensitive to confinement and to surfaces, tiny variations and tiny defects stop being negligible. They become the difference between a sharp spectral line and a blurred smear, between a bright label and a disappearing one, between a display that looks the same on day one and day one thousand.

It helps to think of a dot not as a tiny lamp but as a tiny machine.

A machine can be elegantly designed and still fail in the real world because its parts vary too much, because friction steals energy, or because it wears out. Nanocrystals face the same trio of obstacles, translated into nanoscale terms.

The first battle is uniformity.

When a display engineer says "pure red," they are often really saying "narrow emission." The red primary should occupy a tight region of the spectrum so it does not leak toward orange or deep infrared. That spectral narrowness is not only an optical property of a single dot. It is also a statistical property of a crowd.

A television backlight film contains an astronomical number of particles. If those particles differ in size, even slightly, each size emits at a slightly different wavelength. The film's overall emission becomes a blend: a peak with fat shoulders, a colour with softened edges. The same problem appears in a biological assay. If the label intended to mark one protein actually emits a range of greens and yellow-greens, it becomes harder to separate that signal from other labels, and harder to quantify.

Uniformity has a technical name: *monodispersity*, meaning the particles in a batch are close to one another in size (and ideally shape). A monodisperse sample gives a narrow emission linewidth and crisp absorption features. A polydisperse sample gives broad features. The ruler becomes fuzzy.

The crucial point is that the tolerances are brutal. On the human scale, a few percent variation in size feels like nothing. On the quantum scale, a few percent can shift emission enough to be visible, because the energies depend strongly on size. If a synthesis produces dots with a wide distribution, the very phenomenon that makes them tunable also makes them unforgiving.

This is why the synthetic breakthrough associated with Murray, Norris, and Bawendi (1993) mattered so much to the world outside a chemistry lab. Their hot-injection approach was designed to make nucleation happen in a short burst and then stop, so that growth could proceed mainly by enlarging the existing population rather than continually creating new, smaller particles. A short nucleation window is like a starting gun in a race: everyone begins at nearly the same time. If latecomers keep joining, the pack spreads out. By turning nucleation into an event rather than a long process, the chemistry made it possible to create much tighter size distributions, and therefore much narrower optical features.

Uniformity also has an emotional side that laboratories know well. When a batch is truly monodisperse, the absorption spectrum can show distinct "steps" or peaks rather than a smooth blur, and the emission looks clean. There is a palpable sense that the material is behaving like a designed object rather than an uncontrolled mixture. That feeling is part of why quantum dots became a platform rather than a one-off curiosity: controllability invites ambition.

The second battle is brightness.

Brightness is not simply "how many photons come out" in some absolute sense. It is fundamentally about *efficiency*: how much of the absorbed energy returns as light rather than disappearing into heat or other loss channels. The standard metric is the photoluminescence quantum yield, the fraction of absorbed photons that re-emerge as emitted photons. A high quantum yield means the material is doing what you want most of the time. A low quantum yield means you are feeding energy into an expensive dead end.

At the nanoscale, brightness is often decided at the surface.

A perfect bulk crystal has a periodic lattice that supports clean bands. A surface has broken periodicity, dangling bonds, reconstruction, adsorbed molecules, and sometimes outright chemical damage. These imperfections can create electronic states inside the band gap—states that can trap charge carriers and provide nonradiative pathways. An excited electron and hole are supposed to meet and emit a photon. A trap state offers a shortcut: energy can be lost without light, often as vibrations (heat). The dot still absorbs, but it does not repay the energy in photons.

This is why surface chemistry is not a finishing touch; it is a core part of optical design. The molecules bound to the surface (ligands) can either passivate dangling bonds—essentially covering up electronic "loose ends"—or leave traps exposed. They can also alter the local electric fields and subtly change energies. In the language of the lab, surface engineering can rescue a dim sample and turn it bright without changing the core size at all.

A vivid, practical image helps here. Imagine a concert hall where sound should echo cleanly. Now punch a hundred small holes in the walls and fill them with thick curtains. The music is still played, but the sound leaks into the walls and dies. Surface traps are those holes. Passivation is the repair crew.

One widely used strategy is to grow a shell of a wider-band-gap semiconductor around the core. The shell functions like a protective coat: it can reduce the influence of surface defects on the core's electronic states and confine the electron and hole away from the outermost surface where traps are most likely. The details can be sophisticated—shell thickness, band alignment, alloying, graded interfaces—but the intuition is simple: keep the fragile, light-emitting core from directly touching the messy outside world.

Brightness is also where the seduction of ensemble measurements can mislead. A vial can look reasonably bright on average while hiding unruly behaviour at the single-particle level. That was made uncomfortably clear when researchers began to watch individual dots rather than crowds. In 1996, Nirmal, Dabbousi, and colleagues reported fluorescence intermittency in single cadmium selenide nanocrystals: individual dots blinked on and off under steady illumination. In a large ensemble, those on/off events average out into a steady glow. For applications that depend on reliable emission—single-particle tracking in cells, for example, or certain device architectures—that intermittency matters.

Blinking is not merely an oddity. It is a symptom of how easily a nanocrystal's emissive behaviour can be hijacked by charging events and surface-related processes. If a dot becomes charged, it can enter regimes where nonradiative recombination is more likely, and emission is suppressed. Some designs and shells can reduce blinking by stabilizing charges and smoothing interfaces; others can make it worse. The discovery of blinking turned "brightness" from a static property into a dynamic one: not only "How bright is it?" but "How reliably does it stay bright?"

The third battle is stability.

A dot that emits beautifully on Monday and poorly on Friday is not a technology; it is a laboratory anecdote. Stability comes in many forms: resistance to oxygen and moisture, resistance to heat during processing, resistance to photodegradation under intense illumination, resistance to chemical attack during ligand exchange and film formation, and resistance to slow spectral shifts that move the emission peak over time.

The causes are similarly varied. Surfaces can oxidize. Ligands can detach. Dots can sinter or fuse when heated, effectively growing larger and changing colour. In a solid film, particles can interact with one another or with the matrix in ways they did not in solution. Even without catastrophic failure, subtle changes accumulate: a small redshift in emission, a broadened linewidth, a drop in quantum yield.

Stability also has a less obvious enemy: the very brightness that devices demand.

Under gentle excitation, a dot typically contains one electron-hole pair at a time. Under intense excitation—the sort used in high-brightness LEDs, lasers, or certain imaging conditions—a dot can host multiple carriers. In that multicarrier regime, an additional nonradiative pathway can dominate: Auger recombination. Instead of emitting a photon, the recombination energy is given to a third carrier, which is kicked to a higher energy and then thermalizes. The result is lost light and added heat, and it can happen on ultrafast timescales. This process is not a rare corner case; it is a central limitation when pushing nanocrystal emitters toward high brightness and high current densities. Shells and heterostructures can be designed to suppress Auger recombination, but doing so is part of the engineering challenge, not a guaranteed benefit of "being quantum."

The stability battle is therefore not only about surviving the environment; it is also about surviving the operating conditions of the intended application. A label used in a low-light microscope experiment faces different stresses than a backlight film sitting behind an LCD, and a different set again from an emissive device where current is driven directly through the dot layer. The same nanocrystal core can be a star in one setting and a failure in another because the dominant loss mechanisms change.

Threaded through all three battles is a constraint that does not come from physics but still shapes the technology: regulatory acceptability and safety.

Some of the best-performing nanocrystal systems have historically been cadmium-based, such as CdSe. Cadmium raises legitimate environmental and health concerns, and consumer electronics are governed by restrictions such as RoHS in Europe. Exemptions have sometimes been granted for specific uses and configurations when the amounts are tightly limited and when efficiency benefits can be argued, but the pressure has been steady: reduce or eliminate cadmium where possible. That has driven intense work on cadmium-free alternatives.

Indium phosphide (InP) dots, for example, reduce heavy-metal concerns but often struggle to match CdSe in photoluminescence quantum yield, emission linewidth, and long-term stability. Perovskite nanocrystals can deliver exceptional colour purity and brightness, but they raise their own stability challenges and often involve lead, reintroducing toxicity concerns in a different form. These trade-offs are not a footnote. They are part of what "a good dot" means in the real world: not only optically excellent, but manufacturable, durable, and acceptable to regulators and to public expectations.

Seen together, the battles explain why the field has been as much chemistry and materials science as it has been quantum physics. The quantum idea provides the knob. Engineering decides whether the knob can be turned precisely, whether it stays where you set it, and whether the light you paid for arrives on cue.

A display that makes a convincing red is therefore not simply a triumph of colour science. It is evidence that a vast number of nanocrystals have been made nearly the same size, given surfaces that favour light emission, protected against the abuses of processing and operation, and packaged in a way that survives both physics and bureaucracy. The reward for winning those battles is a material that behaves less like a fragile laboratory specimen and more like a dependable component: a nanometre-scale object that can be shipped, laminated, heated, illuminated, and still keep its promise to glow.

Chapter 2

The 'Particle-in-a-Box' Hidden in Plain Sight

Why should making a crystal smaller change the color of light it absorbs and emits? The answer lies in a quiet quantum rule hiding inside ordinary matter: once electrons are squeezed into nanoscopic spaces, geometry begins to dictate energy, and color becomes a measure of confinement.

2.1 Waves in a Room with No Escape

A useful superstition clings to the way many of us picture electrons: tiny hard beads that zip around like microscopic billiard balls. That picture is not *wrong* so much as it is disastrously incomplete whenever the space available to the electron becomes small enough that the boundaries start to matter. Then the better mental model is a ripple—a wave—that must live somewhere. And a wave that has to live inside a confined region cannot take just any shape. It must *fit*.

The "fit" is not a vague metaphor. Everyone has met it in ordinary life, usually without naming it. Run a wet finger around the rim of a wine glass and you can coax a clear, single note out of it. The glass does not sing at arbitrary pitches; it chooses particular ones, because the vibration has to wrap around the rim and meet itself smoothly.

Pluck a guitar string and it does not produce a continuous smear of tones. The string produces a fundamental and a ladder of harmonics, because a string fixed at both ends must have nodes—points that remain still—right where it is clamped. Those fixed ends enforce a strict rule: the wave on the string must be zero at the ends, which only certain standing-wave patterns can satisfy.

Electrons, when confined, behave in the same spirit. The walls do not have to be literal; a "wall" can be an abrupt change in material, an electrostatic barrier, or simply an energetic penalty for being in the wrong place. But once an electron is effectively trapped, the question becomes: what wave patterns are allowed inside the trap?

The remarkable step is to stop thinking of confinement as merely stopping a marble from escaping, and start thinking of confinement as an instruction given to a wave: *you may exist here, but you must vanish or die away there.* In quantum mechanics this instruction is called a *boundary condition.* It is a recipe for turning a continuum into a set of discrete possibilities.

It is also the quiet reason that "energy levels" are not an arbitrary quantum quirk. Discreteness is what happens when a wave is forced to live in a finite space under strict boundary rules. Put differently: quantization can emerge because geometry is picky.

To see why, take the simplest imaginable trap: a one-dimensional corridor of length L, with impenetrable doors at both ends. A wave confined to this corridor must be zero at each door. The wave cannot end on a nonzero value; there is nowhere for it to go. The only stationary patterns that satisfy that are the ones that form standing waves: patterns that look like a sine curve with one, two, three, … half-wavelengths fitting exactly between the doors. The simplest pattern bulges in the middle and touches zero at both ends: one half-wavelength. The next has an extra node in the middle: two half-wavelengths. Each additional pattern adds another node and another half-wavelength.

Nodes are more than a mathematical detail. A node is a forced stillness. If you clamp a string, the clamp *is* a node. The wave is not allowed to "negotiate" with it. In the electron's case, the walls act like clamps for its wave. That is why the allowed patterns are countable: there is a first, a second, a third. Between them there is no room for a pattern that almost fits, in the same way there is no guitar note that is *almost* the second harmonic. A wave either satisfies the boundary conditions everywhere or it does not represent a stable, stationary state of the confined particle.

Once the allowed wave patterns become countable, the energies become countable too. The reason is that the energy of a confined quantum particle is tied to how rapidly its wave wiggles in space. A gently varying wave corresponds, in a rough but reliable sense, to a lower kinetic energy; a rapidly oscillating wave corresponds to a higher kinetic energy. The ground state in a trap is therefore not a flat, featureless "rest" state, because a flat state would not go to zero at the boundaries. The lowest-energy wave already has curvature: it must rise from zero at one wall, reach some maximum, and return to zero at the other wall. Confinement does not merely organize energies into steps; it *pushes* the lowest step upward. There is no way to trap a particle and also allow it to have exactly zero kinetic energy, because a perfectly uniform wave cannot respect the walls.

Now comes the size effect that ultimately makes nanocrystals glow in different colors. When you shrink the corridor—when L becomes smaller—it becomes harder to fit a long wavelength inside. The smallest allowed pattern still has to be zero at the ends, but now it must accomplish that within a shorter distance, so it must bend more sharply. Sharper bending means higher kinetic energy. And because all the patterns are squeezed, the gap in energy between successive allowed states also grows. A small box does not merely shift all energies up; it spreads the rungs farther apart.

This is a deeper idea than it first appears. It says that discreteness is not something sprinkled onto nature like a rule in a game. It is a consequence of allowing waves to exist only under particular constraints. The constraints may come from walls, from material interfaces, from electric fields, from the structure of a molecule, or from the edges of a tiny crystal. The mathematics changes in detail, but the logic remains: the space dictates the allowed patterns; the patterns dictate the energies.

A historical anecdote makes this "wave in a room" intuition feel less like abstract philosophy and more like engineering. In the late nineteenth century, when J. J. Thomson and others were still working out what electrons were, the prevailing mental pictures were mechanical. But by the 1920s, after Louis de Broglie proposed that matter has a wavelength and after Erwin Schrödinger wrote down his wave equation, the "room" metaphor became literal enough to calculate with. Schrödinger did not begin with a mysterious decree that energy must come in chunks; he began with a wave equation and asked what happens when a wave is confined. Out of that came discrete energies, in the same way that a violin string gives discrete resonances. Quantization arrived as a boundary-value problem.

The idea quickly became practical. Physicists learned to treat the electron in certain solids as if it were freer than in an atom, yet still restricted by the solid's structure. A crystal can look like a periodic landscape of allowed and forbidden regions, and electrons in it can often be described as waves that extend over many atoms. Then, if you carve out a small region where an electron is more comfortable—lower in energy—you have built an electronic "room." In modern language the simplest version of that room is called a *quantum well* (a trap in one dimension), and the three-dimensional cousin is a *quan-*

tum dot. The names sound futuristic; the underlying principle is the same as the guitar string.

It is worth pausing on a common misunderstanding. When people hear "the electron is a wave," they sometimes imagine a kind of smeared-out jelly of charge. The more accurate picture is a wave of *probability amplitude*—a mathematical object that tells you where the electron is likely to be found if you look. The wave can interfere with itself, create nodes, and form standing patterns. That is why confinement matters so much: nodes become enforced absences. In a trapped state, there are places inside the room where the electron is simply never found, not because it is repelled by some force there, but because the allowed wave pattern is zero there. A node is a *forbidden location* created purely by wave geometry.

This also explains why the energy ladder does not behave like the shelves of a bookcase where the spacing is the same. In the idealized textbook box, the energies spread out faster as you go up the ladder: the higher patterns wiggle more, and each additional wiggle costs more energy. One does not need the formula to get the message: squeezing a wave costs energy, and squeezing it harder costs disproportionately more.

At this point, it helps to hold two boxes in mind: one large, one small. In the large box, the lowest standing wave is long and smooth. The next few patterns are not dramatically different, and the energy differences between them are relatively modest. In the small box, even the lowest pattern has to contort to satisfy the boundaries. The second pattern contorts more, and the energy gap between the two is no longer a subtle correction; it is a dominant feature of the system's behavior.

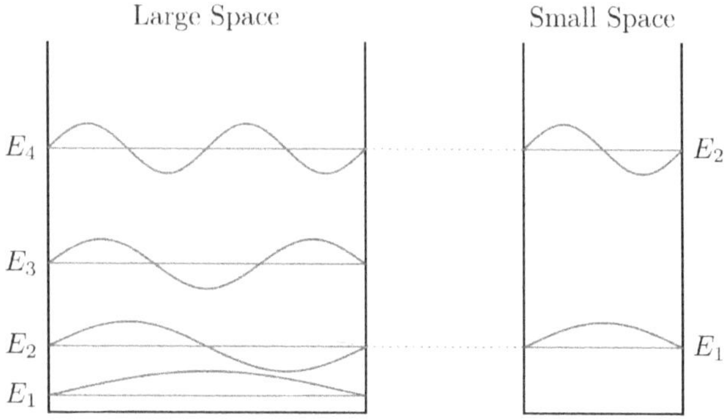

Fewer long-wavelength standing patterns fit inside a smaller space, so the lowest allowed state is higher and the spacing between allowed energies is larger.

The box model is, of course, an idealization. Real materials do not build infinitely hard walls; their boundaries are softer, and an electron's wave usually leaks a little into the surrounding region. If you trap sound in a room, the walls are never perfect either: low frequencies can pass through, and a little energy always seeps into the building. Yet rooms still have resonances. The resonances broaden, shift, and become less "pure" as the walls soften, but they do not vanish. Likewise, a trapped electron in a realistic potential does not have perfectly sharp boundaries, but it still has a discrete set of allowed stationary states as long as it remains meaningfully confined.

This is where the analogy earns its keep. The value of the particle-in-a-box is not that any real electron lives in a perfectly square room with infinitely high walls. The value is that it captures an unavoidable principle: *confinement converts free motion into quantized patterns, and smaller confinement makes the quantization more severe.* When engineers and chemists later learned to make tiny crystalline regions inside a different material, or to grow nanoscale crystals suspended in a liquid, they were not importing an abstract textbook toy into the lab. They were discovering that matter itself can provide the walls.

An electron moving through a large, bulk crystal is often described by band theory: it can take on a near-continuum of energies within allowed bands, because the electron's wave can extend over a huge number of atoms and the spacing between allowed wave patterns becomes so tiny that it effectively disappears. In a macroscopic chunk of semiconductor, the "room" is so big that the rungs of the energy ladder are nearly touching. That is why bulk materials feel continuous.

But shrink the crystal down to nanometre scales and the continuity becomes an illusion you can no longer sustain. The electron's wave now has to "make do" with a much smaller region. The boundaries are real: the crystal ends, the material changes, the allowed electronic states change sharply. The electron's wave begins to resemble the standing waves of the box. Energies near the band edge, instead of forming a near-continuum, begin to separate into distinct levels. The ladder is no longer a fine-toothed comb; it starts to look like steps.

A concrete example anchors this in something you have likely seen. The glow of a traditional fluorescent lamp comes from atoms in a gas, and the color is fixed by the atomic species. By contrast, the modern, richly saturated reds and greens in many high-end televisions and tablets often come from nanocrystals embedded in a film. Those nanocrystals are made of semiconductors whose bulk form would have a fixed optical character. Yet in nanoscale form they can be tuned: one batch emits a warmer red, another a yellow-green, another a crisp green. That tuning does not come primarily from dye chemistry, as in a pigment, but from how tightly electronic waves are confined. The "room" has been resized.

The key is that the energy scale shifts dramatically with size. The electron is light; it does not take much to change the energy cost of forcing its wave to have an extra node or to wiggle faster. Nanometres are not merely small in the everyday sense. They are comparable to the natural lengths associated with electrons in solids. When the confining dimensions approach those lengths, the boundary conditions stop being a minor detail and start setting the rules of the game.

It is tempting to ask, at this stage, whether quantization requires an electron to be literally bouncing between walls. It does not. The standing wave is not a record of a tiny ball ricocheting. It is a stable pattern for the electron's probability amplitude, a pattern that persists in time except for an overall oscillation that carries the energy. If you measure the electron's position, you will find it at a particular location, and repeated measurements build up the pattern. But between measurements, the wave description is not a story of a particle tracing a definite path; it is a story of allowed and forbidden waveforms.

That difference matters because it makes the "walls" in a material more flexible than a box in a cartoon. A wall can be created by an energy barrier rather than a physical barrier, and an energy barrier can be engineered. Chemists can change it by swapping atoms, coating surfaces, or surrounding a nanocrystal with a different material. Device designers can change it with electric fields. In every case, the wave cares about the boundary conditions, and the boundary conditions determine what "fits."

The particle-in-a-box metaphor has limits that are easy to respect once they are named. First, real nanocrystals are three-dimensional and their shapes are not perfect cubes; they are spheres, rods, plates, truncated polyhedra. That changes the detailed pattern of allowed states, just as a drumhead differs from a string. Second, real boundaries are not infinite; the wave can leak, and leakage blurs and shifts energies. Third, electrons in solids do not behave exactly like free electrons because the crystal lattice and interactions with other electrons modify their effective inertia. None of these caveats destroys the central lesson. Each simply reminds you that "box" is shorthand for "confined region with boundary rules."

If the metaphor is held gently rather than literally, it becomes a powerful piece of intuition you can carry into the nanoscale world. A confined electron is a wave forced to stand. Nodes are enforced absences. Allowed patterns come in a sequence. Shrinking the available space raises the energy of the lowest allowed pattern and widens the spacing between patterns. A small room does not just constrain what happens inside it; it changes what the inside *is allowed to be*.

2.2 Why Small Crystals Glow Blue

Once the "wave in a room" picture is in your head, semiconductors stop behaving like opaque chunks of engineering and start acting like musical instruments for electrons. The key feature of a semiconductor is not that it is halfway between a metal and an insulator—that is a consequence, not a cause. The cause is that its electronic states come in broad families separated by a forbidden zone.

In a typical semiconductor, the lower-energy family is called the *valence band*. At ordinary temperatures, it is largely filled with electrons. Above it sits the *conduction band*, which is mostly empty. Between them is an energy range where the material simply has no allowed electronic states. That forbidden range is the *band gap*. The name makes it sound like a bookkeeping detail, but optically it behaves like a threshold: to move an electron from the valence band into the conduction band, you must supply at least the gap energy. Light can supply it in a single packet—a photon. If the photon is too low in energy, nothing happens; if it is high enough, the semiconductor can absorb it and promote an electron upward.

That is why semiconductors have characteristic colors of absorption and emission in the first place. The gap sets the scale of which photons can be eaten and, when the system relaxes, which photons can be spat back out.

Now imagine shrinking the semiconductor until it is no longer an effectively infinite crystal. In a bulk wafer, an electron near the bottom edge of the conduction band can have a range of energies and momenta so finely spaced that the ladder looks continuous. In a nanocrystal, the electron is not free to roam across billions of atoms. Its wave must fit into a finite region, and the allowed patterns become discrete. The conduction-band edge and valence-band edge stop being smooth ledges and start resembling sets of steps.

The practical consequence is blunt: the effective gap that governs optical transitions grows as the nanocrystal gets smaller. When the gap grows, the photons that match it must carry more energy. More energy means shorter wavelength. Shorter wavelength means bluer light.

That statement can be made without a single equation, but it is not hand-waving. It rests on a piece of physical logic that is as solid as the resonance of a guitar string. Confinement forces the electron's wave to bend more sharply. Sharper bending corresponds to higher kinetic energy. In a semiconductor nanocrystal, there are two bending waves to think about: the promoted electron in the conduction band and the missing electron it leaves behind in the valence band. Each experiences confinement. Each pays an energetic price for being squeezed into a small volume. The optical transition energy is, roughly speaking, the bulk band gap plus the confinement costs.

There is a historical moment when this intuition became a calculable framework rather than a slogan. In the early 1980s, as researchers began to treat small semiconductor crystallites as objects with their own electronic structure, Aleksei Efros & Alexander Efros (1982) provided theoretical grounding for how interband light absorption changes in semiconductor spheres when size becomes comparable to characteristic electronic length scales. Soon after, Louis Brus (1984) wrote down an effective-mass picture that made the size dependence especially transparent. The dominant contribution to the blue shift in that framework comes from quantizing electron and hole motion, giving a confinement term that scales roughly like $1/R^2$, where R is the crystallite radius. There is also a counteracting piece: the electron and hole attract each other, and their Coulomb pull contributes an energetic correction that scales roughly like $1/R$, partly offsetting the blue shift. The two terms fight, but the $1/R^2$ confinement term wins as the crystal becomes small enough.

Those scalings are worth translating into everyday language. A $1/R^2$ term is not a gentle nudge; it is a steep penalty for smallness. Cut the radius in half and this contribution becomes four times larger. The Coulomb attraction also strengthens as the carriers are squeezed closer, but it grows more slowly. So when you shrink a nanocrystal enough, the spacing of the electronic steps expands quickly, and the optical gap increases.

One way to make this feel real is to think in terms of the color wheel that lives inside every screen. A display cannot easily produce a vivid red, green, and blue using white light filtered by dyes without wasting a lot of photons. Modern devices instead aim for narrow, well-placed emission peaks: pure-ish primaries that can be mixed efficiently. Semiconductor nanocrystals became commercially valuable because they offer precisely that: very saturated emission colors that can be tuned to a desired wavelength by adjusting the nanocrystal size during synthesis. The tuning knob is physical size, not a new dye molecule each time.

A common misconception is that the color is a surface stain, as if the nanocrystal were a tiny glass bead tinted by impurities. Another misconception is that the color is a chemistry trick, as if the nanocrystal were a molecular dye whose hue changes because the chemical formula changes. Chemistry does matter enormously—composition determines the baseline band gap and the microscopic details of the electronic states—but the size tuning is a different kind of control. It is closer to changing the pitch of a whistle by changing its length than it is to mixing a new pigment.

It also helps to separate two optical ideas that get conflated: *absorption* and *emission*. Absorption is about what the nanocrystal can take from incoming light. Emission is about what it gives back when it relaxes. In an ideal, perfectly clean semiconductor nanocrystal, the onset of strong absorption occurs near the effective gap, and the emission occurs not far below that energy. In reality there is usually a small offset: emission tends to be slightly red-shifted relative to the absorption edge because the excited carriers lose a little energy to vibrations and settle into the lowest available excited state before recombining. This difference is called the Stokes shift in many contexts. For a reader's intuition, the important point is simple: the same size-dependent electronic structure that shifts absorption also shifts emission. Shrink the crystal and both edges move toward higher energies, which correspond to bluer photons.

A concrete example is the classic family of cadmium selenide nanocrystals, studied for decades because they offer clean, bright emission across much of the visible range when properly prepared. In bulk form, cadmium selenide has a fixed band gap. But as nanocrystals, its emission can be moved from the red toward the green and into the blue as the particles get smaller. The material is the same in the sense that the chemical composition is the same; what changes is the confinement. The electron and hole are being forced into a tighter "room," and the energetic price shows up as a bluer photon.

Even without numbers, you can feel the logic. Red light has lower photon energy than blue light. If a smaller particle emits blue, it must be releasing a larger energy difference between an excited state and a lower state. In a confined system, large energy differences are natural because the allowed standing-wave patterns are farther apart. A large particle has plenty of room; its electronic ladder is crowded, and the relevant transition energy stays closer to the bulk band gap. A smaller particle has fewer low-energy spatial patterns available; the first allowed excited state is pushed upward, widening the gap that light must bridge.

That is the core story, but the honest story has texture. Real nanocrystals are not perfect boxes. They are made of a finite number of atoms with a surface, and surfaces are unruly places. Atoms at a surface are missing neighbors. Bonds are strained or dangling. Charge can leak into surrounding ligands or solvents. All of that creates extra electronic states that do not exist in the pristine bulk crystal. Some of those states act as traps: an excited electron (or hole) can fall into them and lose its energy nonradiatively as heat instead of as a photon. This is why the clean size-color rule is, in practice, also a story about *surface quality*.

This is also why the rule is powerful but not absolute. Size is a dominant knob, yet it is not the only knob. Composition matters because different semiconductors start with different bulk gaps; you cannot get deep red emission from a material whose bulk gap already sits in the blue. Shape matters because confinement in one dimension more than another changes the pattern of energy levels; a rod-like nanocrystal does not confine carriers the same way a nearly spherical one does. The surrounding environment matters because electric fields and dielectric screening change how strongly charges interact and how easily their wavefunctions leak. And surface defects matter because they add alternative pathways that can dim or shift emission.

Still, the size trend is robust because the underlying physics is geometric. Whatever the detailed material, squeezing a wave raises its energy. In Brus's effective-mass framework (1984), the dominant size-dependent term comes from quantizing electron and hole motion and scales like $1/R^2$. The attraction between electron and hole pulls the transition energy downward with a roughly $1/R$ dependence, but it does not erase the confinement effect. Efros & Efros (1982) emphasized that when the characteristic lengths of electronic excitations are comparable to the particle size, you should expect discrete-state effects in absorption. These are not fragile claims. They are the predictable result of waves being forced to live in a small region.

If you want an everyday analogy that does not trivialize the physics, think of a playground swing rather than a paint chip. A long swing has a slow, low-frequency motion; a short swing has a faster motion. The swing's length sets its natural timescale. Confinement sets a natural spatial scale for an electronic wave. That spatial scale sets the energetic spacing between allowed states. The color is the optical expression of that spacing.

There is also an emotional reason this matters: this is one of those places where an abstract rule turns into a design language. When a chemist in a lab controls the growth of nanocrystals by seconds and degrees—stopping a reaction earlier for smaller particles, letting it run longer for larger ones—they are not just changing a size distribution under a microscope. They are sculpting the allowed electronic states, and therefore sculpting the colors a device can generate or a biological label can emit. It is hard to think of a more direct link between nanoscale geometry and something as human as the hue of an image.

The final piece of intuition is to keep "blue" from becoming a mystical label. Blue photons are not special; they are simply more energetic. A nanocrystal that emits blue is not somehow more excited as a personality trait. It is a system whose lowest radiative transition has been pushed to higher energy by confinement. That is why a small semiconductor nanocrystal can be a blue emitter even if a bulk chunk of the same material would never look blue at all. The walls have rewritten the permitted standing waves, and light merely reports the new spacing.

A tiny crystal, then, is not a miniature version of a large one in the way a pebble is a miniature version of a boulder. At nanometre scales, shrinking the object changes the rules that its electrons must obey. The reward for learning to think this way is that the spectrum of a material becomes something you can tune with a ruler.

2.3 A Missing Electron Leaves a Partner Behind

The most misleading sentence in the popular description of semiconductor light emission is also the simplest one: an electron "falls" and a photon comes out. It leaves you picturing a lone particle dropping down an energy staircase into an empty void. Real semiconductors do not offer empty voids. They offer a sea of electrons packed into the valence band, shoulder to shoulder, with the conduction band above it waiting for new arrivals. When light is absorbed strongly enough to matter, the event is inherently a two-character drama.

Absorbing a photon across the band gap promotes an electron to the conduction band and leaves a hole in the valence band; electron + hole can bind via Coulomb attraction to form an exciton. That sentence contains nearly everything needed for the optical physics of nanocrystals, but it hides the intuition. The key is to take seriously what it means to "leave a hole".

A hole is not a little marble of antimatter hiding in the solid. It is the absence of an electron in a band that was almost completely full. Because the valence band is so crowded, a single missing electron stands out the way an empty seat stands out in a full theatre. The remarkable trick of semiconductor physics is that the empty seat can be treated as if it were a mobile, positively charged object in its own right. A *hole* is treated as a positively charged quasiparticle representing the absence of an electron in a nearly full valence band; this is a convenient, physically predictive description for transport and optical phenomena. The word *quasiparticle* is important: it signals that we are not claiming the hole is a fundamental particle like an electron. We are saying that the collective behavior of many electrons in the valence band can be described, cleanly and accurately, as if there were a single positive carrier moving around.

A dinner-party way to see this is to use the "musical chairs" picture, but with the chairs arranged in a lattice. Imagine a long row of chairs, all occupied. If one person stands up, the empty chair is not nailed to the floor as a static emptiness. People can shuffle: a neighbor steps into the empty chair, creating a new empty chair one seat over; then another neighbor steps into that one, and so on. From across the room, what you notice is that the *empty seat* has moved. Nothing physical traveled in the usual sense except a sequence of people shifting position, but the moving absence is a perfectly good thing to track. Better than that: it is the only thing to track if you care about how the gap migrates along the row.

Semiconductors are like that row. The electrons in the valence band can rearrange when one of them is removed. The rearrangement can be captured by saying "the hole moved." This language is not just poetic; it gives the right sign of charge and the right direction of current. If an electron hops to fill a neighboring vacancy, the missing electron appears to have hopped the other way. That is why hole motion often looks like positive charge flowing.

The hole picture is closely tied to band curvature/effective mass: near valence-band maxima, describing dynamics via holes yields positive effective mass behavior. The phrase *effective mass* can sound like a fudge factor, but it is really a summary of how a wavepacket responds to forces inside a periodic crystal. In free space, force gives acceleration in a simple way. In a crystal, the allowed electron waves have a built-in relationship between energy and momentum set by the lattice. Near the bottom of the conduction band, that relationship often resembles a bowl: push the electron and it accelerates as you expect, as if it has a positive inertia. Near the top of the valence band, the relationship resembles an upside-down bowl. Describing the motion there in terms of electrons can lead to awkward signs, as if the electrons had a negative effective mass. The hole description tidies this up: instead of tracking an electron near the top of a filled band, you track the missing electron, and the result behaves like a particle with positive charge and a positive effective mass. The mathematics is precise, but the intuition is straightforward: it is easier to describe the dynamics of the vacancy than the dynamics of every surrounding occupant.

Now return to optics. When a photon is absorbed across the band gap, one electron is promoted into the conduction band. A hole is left in the valence band. You have created a separated pair of opposite charges within the material. Opposite charges attract. Even if the electron and hole do not form a tight, hydrogen-like orbit the way an electron and proton do in empty space, they still influence each other. In many semiconductors, and especially in nanocrystals where the charges are confined to a small volume, that attraction is strong enough that it is useful to treat the electron and hole as a temporary partnership: an *exciton*.

An exciton is not a permanent molecule. It is a correlated state: a conduction-band electron and a valence-band hole moving through the material in a linked way because of their Coulomb pull. The exciton can have its own characteristic size, its own binding energy (the energy you would need to pull the pair apart), and its own ways of interacting with vibrations of the crystal lattice. Optical emission, in many cases, is the annihilation of this partnership: the electron drops back down to fill the hole, and the released energy is emitted as a photon.

This paired view immediately clarifies why the earlier confinement story never quite tells the whole truth on its own. Confinement pushes the electron and hole energy levels around because it quantizes their motion. But the electron and hole also attract each other, which shifts the transition energy in the opposite direction. That tug-of-war is not a small correction; it can shape the spectrum, change the radiative lifetime, and influence how efficiently light is emitted rather than lost as heat.

A historical note helps here, because excitons were not invented for nanocrystals. They were proposed to explain optical behavior in bulk crystals long before nanoscale synthesis became routine. In the early decades of the twentieth century, researchers studying the absorption edges of ionic crystals and molecular solids noticed features that did not look like simple promotion of independent electrons. Yakov Frenkel introduced an early exciton concept in the 1930s to describe bound electronic excitations in crystals where the electron and hole remain close, almost localized on a molecular site. Later, Gregory Wannier and Nevill Mott developed a complementary picture for semiconductors where the electron and hole are more spread out, bound weakly and extending over many lattice spacings. The point is not to memorize names; it is to see that the electron–hole partnership is a deep part of semiconductor optics, not a special trick reserved for nanostructures.

What does change at the nanoscale is how cramped the partnership becomes. In a bulk semiconductor, an exciton can be quite large, spanning many unit cells, because the electron and hole are screened by the material's polarizability and can roam. In a nanocrystal, the boundaries do not allow such a leisurely arrangement. The electron's wave cannot spread indefinitely, and neither can the hole's. The pair is compressed into a finite region, which tends to increase their overlap—how much their probability clouds sit on top of each other. Overlap matters because radiative emission requires the electron to find the hole. A greater overlap often means a higher chance per unit time of recombination that produces a photon, which can shorten the radiative lifetime and increase brightness, provided nonradiative traps do not steal the energy first.

The exciton picture also makes it easier to talk about why absorption and emission lines can be sharp. If you imagined a lone electron making arbitrary transitions, you might expect a smeared continuum. Instead, the correlated electron–hole states can form a set of discrete excitonic levels, particularly in confined systems. Light couples strongly to certain transitions and weakly to others, producing recognizable peaks. This selectivity is one of the reasons semiconductor nanocrystals can act as such clean color emitters when their surfaces are well-passivated.

It is also a way to understand a subtlety about the energy of emitted light. The naive "electron falls" story would say the photon energy equals the band gap (or, in confined systems, the effective gap). The exciton story adds that the emitted photon energy is reduced by the exciton binding energy, because the bound pair sits slightly lower in energy than a completely separated electron and hole. The binding energy is the energetic reward for letting opposite charges snuggle closer. In bulk semiconductors, this reward may be modest because screening weakens the attraction and the exciton is spread out. In smaller structures, screening can be reduced and spatial confinement can enhance interaction, so the binding energy can become more noticeable. Even when confinement makes the effective gap larger overall, the Coulomb attraction pulls the emission energy back down a bit—a built-in counterweight.

This paired viewpoint also clears up a lingering conceptual discomfort: when a photon is emitted, what exactly is the electron dropping into? It is not dropping into nothingness. It is filling a specific vacancy in a nearly full band. The existence of that vacancy is what makes the transition possible. Without a hole, the valence band would be fully occupied and the Pauli exclusion principle would block the electron from returning. The hole is not a mere bookkeeping device; it is the destination.

The hole can be created in more than one way, which matters in practice. Direct optical absorption across the gap makes an electron–hole pair in one step. Electrical injection, as in light-emitting devices, can also create the pair by pushing electrons into the conduction band from one side and pulling electrons out of the valence band from the other, leaving holes behind. In either case, light emission ultimately comes from recombination: electron meets hole, the pair annihilates, and energy exits as a photon (if the system is kind) or as heat (if traps or vibrations grab it).

An everyday consequence shows up in the way nanocrystal emitters can be bright yet also finicky. Bright emission requires that excitons recombine radiatively before they are lost by nonradiative pathways. That sets up a race. The electron and hole feel a mutual pull that encourages them to meet; confinement can enhance their overlap; but surfaces and defects can offer alternative "meeting places" where the energy is dissipated without light. Good nanocrystal chemistry often looks, from far away, like a quest for purity and protection: keep the exciton's energy from leaking into silent channels so that recombination produces photons reliably. The moment you take excitons seriously, this chemical story becomes less mysterious. You are not merely polishing a tiny crystal; you are protecting a delicate correlated state.

The exciton picture also gives a more humane understanding of what it means to "tune" a nanocrystal's color. Tuning is often described as shifting an energy difference, which can feel abstract. In the exciton language, tuning changes the landscape in which this electron–hole partnership lives: the size of the room that confines them, the strength of their mutual attraction, and the degree to which their wavefunctions overlap. Those changes determine both the photon energy and the likelihood that a photon appears at all.

There is a final intuition that is easy to miss: an exciton is not merely two charges orbiting each other; it is also a way for the crystal to carry excitation without transporting net charge. An electron and a hole together are electrically neutral overall. That neutrality helps explain why optical excitations can move through a material in ways that resemble energy transport rather than current flow. In nanoscale systems, where the exciton may be confined to one particle or hop between particles, this distinction becomes a design consideration: are you trying to move charges for electricity, or excitons for light?

Seen this way, the semiconductor does not emit light because an isolated electron gets tired and falls down. It emits light because an electron has been lifted out of a crowded band, leaving behind a mobile vacancy that behaves like a positive carrier, and the two are drawn together into a correlated pair. When recombination happens in the right way, the partnership ends by releasing a photon whose color carries the fingerprint of confinement and interaction, a tiny burst of information about how waves, boundaries, and attraction share the work inside a crystal only a few nanometres across.

2.4 The Size Line Where Quantum Rules Take Over

A nanometre is a tiny length by any human measure, but a semiconductor does not judge size by human standards. It judges size by comparing it to its own internal yardsticks: how far its charge carriers can roam, how strongly they feel each other, how effectively the surrounding crystal screens electric fields. Two crystals can share the same diameter and still live in different physical worlds, because one material's "natural" electron-hole pair is compact while another's is sprawling. The difference is captured by a single, surprisingly practical concept: the *exciton Bohr radius.*

The exciton Bohr radius is a practical characteristic length scale for electronic excitations in bulk semiconductors; when a nanocrystal dimension is comparable to or smaller than it, energies become strongly size-dependent (quantum-size effect). The name is borrowed from the Bohr radius of the hydrogen atom, and the reason is more than historical decoration. An exciton in a semiconductor is often a hydrogen-like problem in disguise: a negatively charged electron and a positively charged hole attract each other, but the attraction is softened by the material's ability to screen electric fields, and the motion is altered because the carriers do not behave like free particles. Put those two modifications together and you get a "hydrogen atom" whose typical size can be much larger (or sometimes smaller) than the one in vacuum. That typical size is the exciton Bohr radius.

Thinking in terms of this radius does something liberating. It replaces the vague claim "small crystals behave differently" with a sharp comparison: *small compared to what?* Compared to the exciton Bohr radius, the size at which an electron-hole pair would like to spread out in the bulk material if you let it. If the crystal is much larger than that, the exciton can form and move around without constantly bumping into boundaries. If the crystal is comparable in size, the boundaries begin to squeeze the exciton. If the crystal is smaller still, the boundaries dominate the physics: the exciton cannot adopt its preferred bulk size, and the energy spectrum becomes strongly size-dependent.

This comparison also gives a satisfying reason why the same nanometre number can mean different things in different materials. The exciton Bohr radius depends on two material properties in particular. First, how "heavy" the electron and hole act inside the crystal, a notion captured by effective mass. Heavier carriers are harder to confine into a small wavepacket without paying energy; they also move more sluggishly, which affects the balance between kinetic energy and Coulomb attraction in the exciton. Second, how strongly the material screens electric fields, set by its dielectric environment. Strong screening weakens the electron-hole attraction, allowing the pair to spread out more. We do not need the formula to see the direction: weak attraction and light carriers tend to give a large exciton, while strong attraction and heavy carriers tend to give a small one.

Once the exciton Bohr radius is in hand, people in the field often talk about three qualitative regimes of confinement. The terms are widely used because they are useful rather than because nature draws crisp borders.

If the nanocrystal radius R is much larger than the exciton Bohr radius, the system is in *weak confinement.* The exciton comfortably fits inside the crystal. Its internal structure remains close to the bulk exciton, and the dominant size effect is not the quantization of individual electron and hole motion but the more subtle influence of boundaries on the exciton as a whole. In this regime, the optical gap is only weakly size-dependent. The system may still show some finite-size signatures, but they are modest corrections on top of bulk-like behavior.

When R is comparable to the exciton Bohr radius, the system sits in an *intermediate confinement* regime. The exciton begins to feel cramped. Confinement and Coulomb attraction compete on roughly equal footing. Energy levels become more discrete, and optical properties become more tunable with size, but the simple extremes do not apply cleanly. This is often the regime where intuition must remain flexible: some materials and shapes show strong size dependence in certain transitions while still retaining bulk-like features in others.

When R is much smaller than the exciton Bohr radius, the system enters *strong confinement*. The boundaries impose themselves aggressively. The electron and the hole are each forced into quantized spatial patterns, and the energy cost of confinement becomes large compared to the Coulomb binding that would hold them together in the bulk. In this limit, it can even make sense to think of the electron and hole as being confined separately, with their mutual attraction treated as a correction. The result is an electronic structure that can resemble a set of discrete, molecule-like levels rather than the quasi-continuous bands of a bulk crystal.

This regime language did not appear by accident. Efros & Efros (1982), in laying early theoretical foundations for size-quantization effects in semiconductor spheres, emphasized that behavior changes depending on how the particle size compares to the characteristic Bohr radii associated with the carriers. The goal was not to invent jargon but to provide a map: the same basic ingredients are present—confinement, Coulomb attraction, screening—but their relative importance shifts with size, and the spectrum shifts with them.

A useful way to make this feel concrete is to imagine trying to keep a dance partnership inside increasingly smaller rooms. In a ballroom, the pair can maintain its natural spacing and rhythm; the walls are almost irrelevant. In a small studio, the walls begin to dictate the choreography: turns tighten, steps become more constrained, and the partnership changes. In a closet, the walls dominate; the pair can no longer dance in the usual way at all. The exciton Bohr radius is the ballroom's preferred spacing expressed as a physical length.

This also connects to a question that arises the moment one hears that size tunes color: does the tuning continue forever as the crystal gets bigger? It cannot. As nanocrystals grow, their electronic structure transitions from discrete, molecule-like levels toward bulk-like (quasi-)band behavior; confinement effects weaken with increasing size. In the large-size limit, the spacing between quantized levels becomes so fine that it effectively disappears at room temperature and under typical optical broadening. The optical gap approaches the bulk band gap, and further increases in size make little difference. Size tunability is therefore most dramatic when the crystal is in, or approaching, strong confinement relative to the exciton scale of that material.

The same logic explains why ultrasmall crystals can behave in ways that surprise even people comfortable with bulk semiconductors. A bulk semiconductor has a well-defined band structure, with the valence and conduction bands forming broad continua of states. In a strongly confined nanocrystal, the very idea of a smooth band can become a convenient approximation rather than a literal truth. The states near the band edges split into separate levels; selection rules and symmetry begin to matter in ways that resemble molecular spectroscopy; the surface becomes a larger fraction of the system and can reshape the wavefunctions. The more you shrink, the more the material begins to feel like a finite object with its own discrete spectrum rather than a fragment of an infinite periodic lattice.

A historical anecdote captures the spirit of this crossover. In atomic physics, the existence of discrete spectral lines was one of the crucial clues that something wave-like and quantized was happening inside atoms. In solids, the early triumph of band theory was almost the opposite: it explained why bulk materials do *not* behave like collections of isolated atoms with sharp lines, because the electronic states spread over the crystal and form bands. Nanocrystals sit between these worlds. They are made of many atoms and retain the chemical identity of a solid, but their finite size revives discreteness. In a sense, they reintroduce atomic-style quantization into a material that, in bulk, seems to have washed it away.

That "between worlds" position is precisely why the exciton Bohr radius is such a helpful ruler. It is a bulk property—defined in the bulk material—used to predict when a finite object carved from that material will cease to behave in a bulk-like manner. It tells you when the boundaries are not merely trimming the edges of a wavefunction but actively rewriting which spatial patterns are possible.

The material dependence of this ruler matters in practice. Consider two hypothetical semiconductors. In one, screening is strong and carriers are light, so the exciton in the bulk is large. In that material, a nanocrystal of a few nanometres can already be smaller than the exciton Bohr radius, pushing it into strong confinement. The optical gap will then be highly sensitive to size, and modest changes in diameter can produce conspicuous shifts in absorption and emission. In the other material, screening is weaker and carriers act heavier, making the bulk exciton more compact. The same few-nanometre nanocrystal could then be comparable to, or larger than, the exciton radius, living in intermediate or weak confinement. Its optical properties will still be nanoscale, but size changes may have subtler effects; other knobs such as shape, surface chemistry, or strain might compete more strongly.

This is one reason that designing nanocrystal emitters is never just about "make it 4 nm." The same nominal diameter can land you on different sides of the size line depending on composition and environment. Even the surrounding medium can matter. A nanocrystal embedded in a high-dielectric matrix screens charges differently than one surrounded by low-dielectric organic ligands. That changes how strongly the electron and hole attract, and it can shift the balance between confinement and Coulomb interaction. In other words, the exciton Bohr radius is not a mystical constant; it is a material-and-environment-dependent scale that reflects real electrostatics.

It also provides a clean way to interpret the quality of size tunability that chemists and engineers care about. In strong confinement, size is a powerful lever because changing R directly changes the quantization energies. That typically yields a pronounced shift in the effective optical gap. In weak confinement, changing R mainly changes the boundary's influence on an exciton that is already comfortable, so the shift is smaller and can be masked by broadening and disorder. If a material is chosen so that the exciton is large in bulk, then strong confinement and strong size tunability arrive at comparatively larger particle sizes, which can make synthesis and surface control easier. If the exciton is small in bulk, reaching strong confinement may require extremely small particles, where surfaces and imperfections become more difficult to tame.

The exciton Bohr radius also clarifies why the interplay between confinement and Coulomb attraction can look different as size changes. In strong confinement, the dominant energy cost often comes from squeezing the individual carriers into the limited volume; their mutual attraction is present but is a partial counterweight rather than the main driver. As the nanocrystal grows and confinement weakens, the carriers have more room, the quantization energy shrinks, and the Coulomb interaction can become relatively more prominent in shaping fine spectral features. The same physical interaction is always there; what changes is its leverage.

None of this should be taken to mean that there is a single magic diameter at which the rules abruptly switch. Real nanocrystals come with size distributions, temperature broadening, surface disorder, and often non-spherical shapes. The exciton Bohr radius is best treated as a guide that tells you when you should start expecting large size effects and discrete spectral fingerprints. It is a warning label as much as a prediction: *past this point, bulk intuition becomes unreliable.*

There is also a quiet philosophical payoff. When people first encounter size-tunable emission, it can feel like a clever trick: one material producing multiple colors by mere resizing. The exciton Bohr radius turns the trick into a consequence. The material carries within it a characteristic scale for its excitations. If you carve out a crystal smaller than that scale, you force the excitation to adapt. The light that comes out is the record of that adaptation.

A small crystal glows blue not simply because small things are energetic, but because the confined electron-hole excitation is being compressed beyond its natural bulk extent. The exciton Bohr radius tells you whether the crystal is a spacious home for that excitation or a cramped one, and once it becomes cramped enough, the spectrum stops being a gentle variation on the bulk and becomes an object in its own right, with tunable, discrete character written into its size.

Chapter 3

How a Dot Learns to Glow (and Why It Sometimes Won't)

A quantum dot's glow looks effortless, but it is the visible outcome of a precarious microscopic negotiation. Inside each nanocrystal, energy can be funneled into pure color, lost into defects, or diverted into strange stop-start behaviors—and once readers see those competing routes, the logic behind shells, ligands, and careful materials design becomes impossible to miss.

3.1 From Jolt to Glow: The Hidden Journey of an Absorbed Photon

A glowing nanocrystal looks calm only because the interesting part happens too fast to see.

Imagine a single dot suspended in a liquid or embedded in a plastic film. A beam of blue light hits it. Nothing about the dot suggests drama: it is only a few nanometres across, smaller than most viruses, and yet it can take a photon and give back a different photon with almost insolent reliability. The trick is that the dot does not emit the moment it absorbs. Absorption is an interruption—a jolt that throws the dot into an excited condition—and light emission is one of several ways the system may later settle the score.

The first step is simple to say and subtle to picture. The incoming photon is absorbed only if it matches an allowed jump in energy. In a semiconductor nanocrystal, those allowed energies are shaped by quantum confinement: squeezing charge carriers into a tiny volume forces their energies into discrete, size-dependent levels. That is why a smaller dot tends to have a larger effective band gap and a bluer color. When a photon with enough energy arrives, it can lift an electron-like carrier from a lower-energy set of states into a higher-energy set, leaving behind a hole-like absence in the lower set. Bound together by their mutual attraction, the pair is called an exciton. It is not two independent particles so much as a coupled disturbance of the nanocrystal's electronic sea, localized to a region that is, inconveniently for intuition, about the same size as the dot itself.

That exciton is rarely created at the lowest possible energy. The dot can absorb photons that are well above its band edge, and the first excited condition is often "hot": extra energy is stored as the carriers occupy higher quantized states. If emission happened immediately from that hot configuration, the outgoing photon would tend to be close in energy to the absorbed one. But that is not what photoluminescence usually looks like. Real dots almost always emit after they have cooled internally, and that internal cooling is both the reason the emission color is well defined and the reason absorption and emission spectra are not mirror images.

Cooling here does not mean the dot's temperature drops below its surroundings; it means the carriers shed their excess energy to the dot's vibrations and to the immediate environment. In a bulk crystal, this loss of energy is often described as the emission of phonons—

quantized lattice vibrations. In a nanocrystal the story acquires extra texture: the vibrations are confined too, the surface is close to everything, and the coupling to surrounding molecules or a host matrix can be strong. Early on, some physicists expected that quantum confinement might slow this cooling dramatically. If the energy steps between quantized levels were larger than the typical phonon energy, then a carrier might be "stuck," unable to lose energy in the usual small packets. This idea became known as the phonon bottleneck. It was a sensible prediction, and it made quantum dots sound even more exotic: little boxes that trap not just particles but their heat.

Experiments complicated the picture. Ultrafast spectroscopy—the art of watching matter evolve on femtosecond timescales—showed that carriers in many semiconductor nanocrystals relax to the band edge extremely quickly, often in less than a picosecond. The bottleneck, if it exists at all, is usually leaky. Several mechanisms can help the dot cool: the electron can hand energy to the hole, the pair can couple to surface vibrations, or to fluctuating fields of nearby ligands and solvents, and the confined system can find routes that a neat textbook ladder might miss. The practical result is that, by the time the dot is ready to radiate, the exciton has typically slid down to the lowest-energy emissive state near the band edge.

Only then does the photon's return become likely. The radiative step—emission of a photon—is a quantum transition in the opposite direction: the exciton recombines and disappears, and the energy it carried leaves as light. The emitted photon is lower in energy than the absorbed one because some energy has already been lost to internal relaxation. That energy difference is the Stokes shift. It is one of the simplest signatures that a dot has been given time to settle before it speaks.

Stokes shifts were noticed long before nanocrystals became a technology. George Gabriel Stokes, working in the 1850s, studied the strange behavior of quinine solutions that glowed blue when illuminated with ultraviolet light. He recognized a pattern: the emitted light was always of longer wavelength (lower energy) than the light used to excite it. To nineteenth-century eyes, this was unsettling. How could matter give back light that was *less* energetic than what it received, yet still be called "the same" phenomenon? The modern view makes it feel almost inevitable. Absorption puts energy into a system; the system shuffles that energy among its internal degrees of freedom; emission returns only what remains in the particular form of a photon. Some of the energy goes elsewhere, often irreversibly, as heat.

In a quantum dot, the Stokes shift can be small compared with that of many organic dyes, but it matters enormously in practice. A small shift means the dot absorbs and emits at relatively nearby wavelengths. That is both a blessing and a nuisance. It is a blessing because the dot can be pumped with light not too far from its emission color, convenient for LEDs and displays. It is a nuisance because emitted photons can be reabsorbed by neighboring dots if the absorption spectrum overlaps the emission spectrum too strongly, complicating thick films and concentrated solutions. Engineers care about Stokes shift the way accountants care about transaction fees: tiny differences become consequential when repeated millions of times across a screen or through a centimeter of tissue.

To make these invisible steps feel less like hand-waving, it helps to think in terms of times. Absorption is effectively instantaneous on the scale of atomic motion: the photon is gone in a femtosecond, and the exciton is created. Relaxation from hot states to the band edge is typically sub-picosecond to a few picoseconds. Radiative recombination is leisurely by comparison: nanoseconds are common, long enough that the dot can interact with its surroundings, long enough that competing loss pathways have time to intrude. Photoluminescence, then, is not a single event but a sequence with a vulnerable middle. Absorb, cool, then emit—unless something else happens first.

That "something else" could be a trap at the surface, an impurity, a vibration that steals the energy in a nonradiative way, or an interaction with another excitation. Those complications belong to later parts of the story. For the moment, even in an ideal dot, the existence of relaxation explains a deep everyday observation: shining higher-energy light on a set of dots does not make them emit higher-energy light. A green-emitting dot excited with violet light still emits green, because the dot forgets the details of the photon that created the exciton. Once the carriers have cooled to the band edge, the outgoing photon mostly remembers only the energy separation of that lowest transition—the dot's effective band gap.

That is why emission is such a faithful color tag for size. In a beaker of dots made with a slightly different diameter, the emission color shifts in an orderly way across the spectrum. In the early 1980s, when Alexei Ekimov observed size-dependent optical properties in semiconductor nanocrystals in glass, and Louis Brus soon after explored similar behavior in colloidal particles, the striking point was not merely that nanocrystals could glow. Many things glow. It was that the *color was tunable by size in a predictable way*. That tunability is what later made nanocrystals central to displays and imaging, and it is also what makes the emission spectrum a diagnostic tool. The dot's photoluminescence is not just a pretty outcome; it is a readout of the dot's internal energies.

Spectra are where that readout becomes visible. If one measures absorption as a function of wavelength, the result is often a curve that rises toward shorter wavelengths with bumps or shoulders. Those features correspond to allowed transitions into various quantized states. Emission, by contrast, typically forms a narrower peak centered near the band edge. The two curves overlap but they are not the same shape. Absorption can occur into many higher-lying states; emission tends to come from the lowest emissive state, because relaxation funnels population downwards before radiative recombination has time to occur.

The width of the emission peak, the linewidth, is another piece of information hiding in plain sight. A perfectly identical ensemble of dots, each isolated from any fluctuation, would emit at a single energy, giving an infinitesimally narrow line. Real samples never do. One reason is simple statistics: no synthesis produces dots all exactly the same size. Even a tight size distribution spreads the band-edge energies, and the ensemble emission peak broadens accordingly. Another reason is disorder and environment. Within a single dot, slight irregularities in shape, composition, or surface structure can shift energies. The surrounding medium can fluctuate, shifting the dot's transition energy through local fields and interactions. Some of these effects are *inhomogeneous*: different dots emit at slightly different energies, broadening the ensemble peak. Some are *homogeneous*: even a single dot has a finite linewidth because its excited state has a finite lifetime and because it couples to vibrations and fluctuating environments. The observed emission line is the sum of these blurrings. A fat emission peak does not merely mean "bad color." It can mean "too many sizes," or "too much disorder," or "a noisy environment," and often it means a bit of each.

People often first encounter linewidth as a consumer experience. A phone screen that uses a broad red emitter looks subtly washed out next to one that uses a narrow red. The difference is not brightness; it is purity of hue. Narrow emission peaks allow a display to hit a wider color gamut without throwing away light in filters. That is why, when nanocrystal films began to be used in LCD backlights, their narrow green and red emission were prized: they could turn a white backlight into primaries that were both efficient and saturated. The public story is usually told as "quantum dots make better colors." The more microscopic story is that the dots emit from a well-defined band-edge transition, and that careful synthesis and processing keep the ensemble linewidth tight enough that the eye sees a clean red rather than a reddish smear.

Stokes shift also leaves fingerprints in spectra. If absorption and emission are separated by a modest but real gap, then one can excite the dot at wavelengths where it absorbs strongly and detect emission where the excitation light is less intrusive. In laboratory measurements, that separation can be the difference between a clean signal and a mess of scattered pump light. In biology, it can be the difference between seeing a labeled cell and being blinded by background. Nature is full of scatterers. Skin and tissue throw light in all directions, and short-wavelength light is scattered especially strongly. A fluorophore with a healthy separation between excitation and emission can be easier to detect amid that glare.

A familiar analogy helps without being too cute. Picture a playground slide. Absorption is climbing up quickly. Relaxation is the slide down to the lowest part of the structure. Emission is jumping off with a shout. The shout does not tell you how hard the child climbed; it mostly tells you where the jump happened. A dot excited with ultraviolet can still emit red; the climb was high, but the jump point is fixed. The Stokes shift is the energy lost while sliding, converted into heat in the structure and the air.

Now for the part that surprises people: the dot can often absorb light at energies far above what it emits, yet the emission energy barely budges. That means the dot acts as an energy *funnel*. You can pour energy in at many wavelengths, and it tends to pour out at one. This is one reason quantum dots behave more like atoms than like chunks of bulk material. An atom has discrete levels; excite it high and it rapidly decays to a particular emitting level before radiating. A dot is not an atom, but confinement gives it a family resemblance.

At this point it is tempting to treat absorption as the input and emission as the output of a neat machine. The trouble is that the machine is built from probabilities. Even the radiative step is not guaranteed. When emission *does* happen, the photon emerges in a direction and polarization set by quantum rules and by the dot's environment. It is emitted after a waiting time that is random but statistically characterized by a lifetime. In an ensemble of many dots, that randomness averages into a smooth glow. In single-dot experiments, it becomes a jittery stream of photon clicks.

Even without discussing the ways dots fail, there is already a reason emission can be weaker than expected: the radiative process takes time. A nanosecond is long at the nanoscale. During that pause, the exciton can be perturbed. If the dot is in a crowded film, energy can hop to a neighbor. If the dot sits near a metal surface, emission can be enhanced or quenched depending on distance. If the dot is warm, vibrations can broaden the line and slightly shift the peak. The glow you see is the end result of a competition that began the instant absorption created an exciton.

The best way to keep the story grounded is to return to what a spectrometer actually shows. When you illuminate a sample and record its emission spectrum, you are looking at the accumulated evidence of three stages: (1) which energies the dot can absorb, (2) how efficiently excitations relax to the band edge instead of being derailed, and (3) how consistently the band-edge transition produces photons of a particular energy. A small change in dot size shifts the peak position. A broader distribution of sizes broadens the peak. More disorder in the environment broadens it further. A larger Stokes shift tells you more energy was lost between absorption and emission, often because relaxation found lower-lying emissive states, or because the dot's excited state geometry and its surroundings reorganized before emission occurred. None of these interpretations require a microscope, only careful attention to light.

To see how quickly this way of thinking became practical, consider a problem faced by early nanocrystal chemists: how do you know whether you have made a narrow size distribution without imaging every particle? Transmission electron microscopy can show size, but it is slow and can be unrepresentative. Photoluminescence offered a shortcut. A sharper emission peak was not merely aesthetically pleasing; it was evidence that the synthesis had produced more uniform nanocrystals. In many labs, the first moment of satisfaction after a synthesis was not a micrograph but the sight of a vial glowing in ultraviolet light with a color that looked clean rather than muddy. The human eye became a crude spectrometer, trained to judge linewidth by the purity of color.

That informal practice was later backed by quantitative spectroscopy. The emission peak could be fitted, its full width at half maximum recorded, its Stokes shift measured, and these numbers could be compared across batches. A new ligand, a different temperature profile, a cleaner precursor—each might sharpen the emission or shift it. In this way, the dot's glow became a feedback signal for chemistry. The photon that leaves the dot carries, in its energy and its spread, a report on the nanocrystal's internal order.

A final detail completes the hidden journey. Even when the dot emits from its band edge, it is not obliged to return to the exact same microscopic state it began in. The surroundings may have warmed slightly; molecules may have reoriented; the dot's surface may have shifted a charge. Fluorescence is a cycle, but not a perfectly closed loop. That is why emission spectra can drift over time in imperfect samples, and why stable emitters are engineered rather than assumed. A dot that seems steady in a photograph may, on longer observation, reveal tiny shifts in peak position and linewidth, the optical equivalent of a voice that is mostly in tune but not perfectly fixed.

So the glow that reaches your eye is less like a reflection and more like a decision made after a brief internal negotiation. A photon arrives with a specific energy. The dot takes it, reshapes that energy through relaxation, and then—if the radiative path wins the race—returns a photon stamped with the dot's size-dependent band edge. The Stokes shift marks the energy spent in the reshaping; the linewidth marks how uniform and quiet the dot and its neighbors are. These are small, technical-sounding quantities, but they are also the reason a nanocrystal can behave like a reliable color source in a television and like a precise optical tag in a laboratory: it tends to forget the details of the light that struck it, and to remember, instead, the energy difference that its own confinement has written into it.

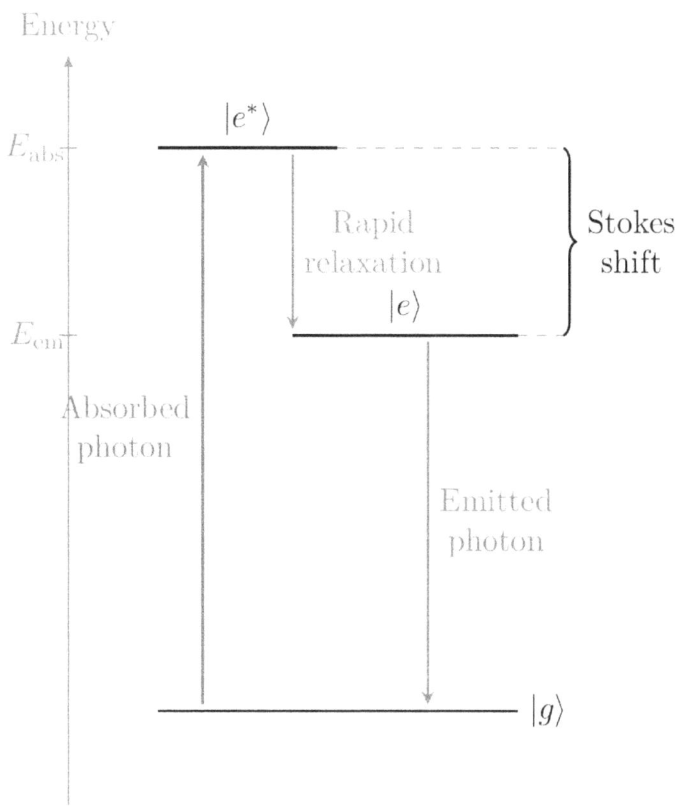

Absorption creates a hot exciton; cooling precedes radiative recombination.

3.2 Brightness Has a Ledger: Counting the Photons You Get Back

Color is the first thing people notice: a vial that glows green, a backlight that looks purer, a microscopic label that stands out from the surrounding haze. Brightness is what decides whether any of that is useful. A vivid hue that is faint might as well be invisible on a sunlit screen, and a perfect spectral fingerprint does not help a surgeon if the signal vanishes under tissue scattering. The uncomfortable truth is that a nanocrystal can be the *right* color and still be a poor light source.

Brightness is bookkeeping. Photons arrive, photons leave, and the difference matters. One part of the ledger asks how often an excitation is created; the other asks how often that excitation produces a photon rather than disappearing silently. The first is governed by the illumination and by the particle's absorption cross section, a per-particle measure of how effectively it intercepts incoming light. The second is governed by the photoluminescence quantum yield, introduced earlier: the fraction of absorbed photons that return as emitted photons. Both parts can vary dramatically between samples that look similar to the eye under casual inspection.

The absorption cross section is an idea that feels abstract until it is tied to a physical situation. Suppose a single dot sits in a beam of blue light. The beam contains a certain photon flux: photons per second per square centimetre. Multiply that flux by the particle's absorption cross section (with units of area), and you obtain an absorption rate: how many photons per second are likely to be absorbed by that one particle. Nothing mystical is hiding here; it is the same logic as estimating how many raindrops hit a window, except the "window" is an effective area that depends on quantum mechanics rather than geometry. For semiconductor nanocrystals, this cross section is strongly size-dependent: larger particles generally present a larger optical "target," so under the same illumination they generate excitations more often.

That size dependence has an immediate consequence that can surprise newcomers: two samples that emit the same color can differ in brightness even if they have the same quantum yield, simply because they absorb differently at the pump wavelength. A deep-red emitter and a green emitter might both be excited with a blue LED, but they can have very different absorption cross sections at that blue wavelength. So when someone says, "This material is brighter," it is worth asking, "Under what illumination?" Brightness is never a property of the emitter alone; it is a relationship between emitter and pumping conditions.

Even when excitation is created, the second part of the ledger decides the payout. Earlier, quantum yield was defined as a ratio. A more revealing way to hold it in your mind is as a race between rates. The exciton can decay radiatively with rate k_r, producing a photon, or it can decay nonradiatively with rate k_nr, producing heat or feeding some loss channel. In that rate picture,

$$\Phi = \frac{k_\mathrm{r}}{k_\mathrm{r} + k_\mathrm{nr}}.$$

The same rates set the photoluminescence lifetime,

$$\tau = \frac{1}{k_\mathrm{r} + k_\mathrm{nr}}.$$

These two compact expressions are powerful because they tell you what brightness changes *mean*. If a dot becomes dimmer while its lifetime stays roughly the same, k_r has likely decreased along with k_nr in a way that preserves their sum. If it becomes dimmer and its lifetime shortens, k_nr has probably increased: the nonradiative escape hatch has widened, pulling population away before a photon can be emitted. Intensity alone is a blunt instrument; intensity paired with lifetime becomes a diagnostic.

A concrete ledger helps. Imagine an illumination that delivers 10^{16} photons per second per square centimetre at the excitation wavelength. Consider two batches of nanocrystals designed to emit the same green. Batch A has an absorption cross section of $10^{-15}\,\mathrm{cm}^2$

at that wavelength; batch B has $3 \times 10^{-16}\,\mathrm{cm}^2$. Even before any discussion of quantum yield, batch A will, on average, absorb about ten photons per second per particle, while batch B will absorb about three. If both batches have a quantum yield of 0.8, batch A emits about eight photons per second per particle and batch B emits about 2.4. To the viewer, batch A looks over three times brighter, but the difference is not "better color" or "better chemistry" in the everyday sense; it is partly simple arithmetic. In real devices, of course, many additional factors enter: how efficiently emitted photons escape a film, how the host material reabsorbs them, how the optics guide them. The point of the exercise is not the exact numbers; it is that brightness can be decomposed cleanly into excitation generation and photon return.

This decomposition also explains a common disappointment in biological imaging. A researcher might purchase a fluorescent label described as "very bright," attach it to antibodies, and then find it barely detectable in cells. Often the label itself is not the culprit; the excitation conditions are. A microscope filter set may excite at a wavelength where the label's absorption is modest, or the illumination intensity may be limited to avoid damaging living tissue. In that regime, absorption cross section becomes decisive. A label with a slightly lower quantum yield but a much higher absorption cross section at the available excitation wavelength can outperform a "better" emitter on paper.

Displays create a different kind of pressure. In a television backlight, the pump is typically a blue LED. The nanocrystals in the film are asked to convert a portion of that blue into narrow green and red. If the particles absorb too weakly at the pump wavelength, the film must be made thicker or more concentrated to capture enough light. But increasing concentration can raise the odds of reabsorption and other optical losses; thicker films can scatter more and waste light in directions the screen cannot use. The engineering goal becomes subtle: maximize useful absorption without creating a self-defeating optical swamp. Once again, "brightness" is a ledger with multiple columns, not a single score.

A historical detour shows why counting photons is harder than it sounds. For much of the twentieth century, fluorescence quantum yield measurements were often relative: compare an unknown sample to a reference dye with a known yield, correct for differences in absorption and detection, and infer the result. This approach works beautifully when the unknown and the standard behave similarly and when optical conditions are gentle. Nanocrystals complicate the situation. They can scatter light, especially in films or in aggregated solutions; they can reabsorb their own emission if the Stokes shift is small; and their absorption can be strong enough that the front of a cuvette is excited much more than the back, producing "inner-filter" distortions. These are not minor experimental nuisances. If you get the bookkeeping wrong, you can claim a quantum yield that is too high by a factor that would make a mediocre sample sound miraculous.

That is why absolute methods became so important, particularly integrating-sphere techniques and de Mello-type approaches, which aim to measure photons-in and photons-out directly rather than leaning on standards. The basic idea is disarmingly physical. An integrating sphere is a hollow sphere with a highly reflective white interior. Light entering the sphere bounces around many times, quickly losing memory of direction. Place a sample inside, excite it, and the sphere allows you to collect essentially all the emitted light, including light that would otherwise be lost to odd angles. By comparing the total light in the sphere with and without the sample, and by carefully separating excitation light from emission light, one can obtain the absolute quantum yield as a true ratio of photons. For nanocrystals—

especially in scattering films and complex matrices—this has been a methodological turning point, because it ties the quantum yield back to the ledger that matters: the actual number of photons that make it out.

Even with absolute methods, one must treat measurements as a craft. Detectors have wavelength-dependent sensitivity; optical components transmit some colors better than others. A sample can heat under illumination, shifting spectra and changing rates. In solutions, concentration matters: too concentrated and reabsorption becomes severe, too dilute and signals drown in noise. Films add their own complications: interference, scattering, and waveguiding can trap light that a detector misses unless the geometry is designed to capture it. When someone quotes a quantum yield without specifying how it was measured and under what conditions, it is like quoting a bank balance without saying the currency.

The rate picture, k_{r} versus k_{nr}, also reveals why two materials of the same color can age differently. Suppose oxygen or moisture creates new nonradiative pathways over time by modifying the surface chemistry. That increases k_{nr}. The quantum yield drops, the lifetime shortens, and the material dims. Another sample might be protected well enough that k_{nr} stays low, but the radiative rate k_{r} changes because the local optical environment changes—perhaps the host polymer slowly rearranges, altering how the emitter couples to the electromagnetic field. Now the quantum yield might not change much, but the lifetime could drift, and the brightness could still move in unexpected ways. These are not academic distinctions. In a medical diagnostic test that must work reliably after months on a shelf, drift of a few tens of percent can determine whether a faint signal is classified as positive or negative.

A simple human-scale example brings the ledger back to lived experience. Picture a surgeon using fluorescence guidance to find the margins of a tumor. The fluorescent tag is not lighting up an empty room; it is competing with the glow and scatter of real tissue. The eye and the camera see a mix of desired signal, reflected excitation light, and background fluorescence from the body itself. Under those conditions, higher quantum yield helps, but so does a large absorption cross section at a safe excitation wavelength, and so does an emission spectrum that falls in a window where tissue is more transparent. Brightness becomes an ethical as well as a technical issue: a brighter tag can mean lower required excitation intensity, reducing photodamage and heating. It can shorten exposure times, reducing motion blur and improving confidence in decisions made in minutes.

The same logic appears in a living room, just dressed differently. A brighter emitter in a display can allow the backlight to run at lower power for the same perceived brightness, extending battery life in a tablet or reducing electricity use in a wall-sized television. It can also improve performance in sunlight, where glare forces the screen to fight the environment. Here the ledger is counted in milliwatts and minutes rather than photons, but the underlying arithmetic is still photons-in and photons-out.

One more subtlety is worth keeping in mind: brightness is often discussed as if it were an intrinsic property of a nanocrystal, like its diameter. Yet the very act of measuring brightness can change the emitter. High excitation intensities can push a dot into multiexciton states, where more than one exciton exists at a time. That regime is important in lasers and in high-brightness applications, but it is also where new loss processes can appear and where simple photon-counting assumptions break. Even before reaching such extremes, excitation can photocharge a particle, alter its local chemistry, or heat its environment. The ledger, in other words, is not always a passive account; the act of spending photons can change the terms of the account.

That said, the ledger framing is liberating because it separates questions that otherwise get tangled. If a sample looks dim, one can ask two clean questions. Is it creating excitations often enough under the available illumination, or is the absorption cross section too small at the pump wavelength? And once excitations are created, are they being converted into emitted photons often enough, or is the nonradiative rate too high? Those questions point to different remedies. One might adjust the excitation wavelength, change film thickness, or select a different emitter with stronger absorption. Or one might improve surface passivation, alter the host environment, or redesign the nanocrystal architecture to suppress nonradiative losses. Without the ledger, all of these interventions blur into a vague desire for "brighter dots."

The most satisfying part of this accounting mindset is that it respects what the eye already knows. When a screen looks washed out, the complaint is not merely aesthetic; it is often a symptom of photons lost to filters, reabsorption, scattering, or low quantum yield. When a fluorescent image is noisy, the problem is often not the camera but the balance between excitation creation and photon return. Photons are stubbornly countable things. They arrive one by one, they leave one by one, and in the space between those events, a nanocrystal decides—through the competition of k_r and k_{nr}—whether it will pay you back in light or keep your energy as heat.

3.3 Where the Trouble Lives: Surfaces, Traps, and Vanishing Light

A nanocrystal's surface is not a skin; it is a large fraction of the body. That single fact changes almost everything about how these particles behave as light sources. In a macroscopic crystal, most atoms live in a comfortably repetitive interior. The surface is a small perturbation that can often be polished, coated, or ignored. Shrink the crystal to a few nanometres and the arithmetic flips. A sizable share of atoms now sit at or near the exterior, where the tidy rules of the interior break down. Bonds end, coordination changes, foreign molecules attach and detach, and the environment can reach in with chemical fingers. The result is that the most important electronic events often happen at the boundary.

The optical consequences can be dramatic. Two samples can be tuned to the same emission color by size, yet one glows brilliantly while the other looks embarrassed, as if it cannot quite manage the task it was designed for. The explanation is rarely in the idealized energy levels; it is in the messy geography of defects. A surface that is chemically perfect in a textbook sense is almost impossible in a flask. Undercoordinated atoms (the so-called dangling bonds), slight non-stoichiometry, missing ligands, and impurities can introduce *trap states*: extra energy levels in or near the band gap that provide a shortcut for losing energy without emitting a photon.

The word "trap" is vivid for a reason. An exciton created by light is meant, in the best case, to relax to the band edge and then recombine radiatively. A trap state offers an alternative landing site. Once a carrier falls into such a state, it may be held there long enough that the clean radiative path is interrupted. The energy can leak away as vibrations (heat), or it can be released in dimmer, red-shifted "trap luminescence" that is broader and less controlled. Either way, the dot's glow is reduced, and its color purity may suffer.

Surface-to-volume ratio is the blunt quantitative driver behind this sensitivity. For a sphere, surface area scales as r^2 while volume scales as r^3, so the ratio goes like $1/r$. Halve the radius and you double the fraction of atoms exposed to the outside world. The intuitive conclusion is that shrinking a crystal makes it more "surface-like," but the practical conclusion is harsher: every small imperfection that would have been negligible in a larger particle becomes a first-order design problem. The same miniaturization that gives tunable color also increases the number of places where light can be stolen.

This is not merely a slow leak that occurs after the exciton reaches the emissive band edge. Ultrafast spectroscopy has made it clear that trapping can compete on very early timescales. In the broader literature, picosecond to tens-of-picosecond windows are often reported for trapping processes. That is an awkwardly short time. The dot can lose a carrier before it has fully thermalized into the lowest excitonic state. In ledger terms, the dot can be robbed before it even has the chance to decide whether to emit. This helps explain a puzzle that used to frustrate newcomers: why an ensemble may absorb light strongly and still emit weakly. The loss can occur so early that the familiar nanosecond radiative lifetime hardly gets a chance.

It also helps explain why the surface chemistry can change not only the *amount* of light but sometimes the character of the emission itself. If trapping is fast, then the spectrum you observe may contain a mixture of band-edge emission and defect-related emission. The latter is often broader because traps are not a single well-defined level; they are a family of states shaped by local chemical disorder. One particle's surface may host a slightly different arrangement of atoms and ligands from its neighbor's, producing a distribution of trap energies and thus a smeared-out emission tail.

A particularly helpful framing, emphasized by researchers including those at NIST, is that fully passivated nanocrystals can avoid surface-derived states in the fundamental gap, while unpassivated dangling bonds can produce surface-state bands near band edges, turning surfaces into efficient loss channels. The phrase "surface-state band" is telling. It suggests that a poorly terminated surface is not just a handful of isolated defects; it can be a dense collection of states that behave almost like a separate material glued onto the nanocrystal. If the band-edge wavefunctions overlap that region, the carriers can be siphoned away repeatedly. The dot is then not simply "a good emitter with a few defects." It has, in effect, grown an electronic sponge on its outside.

Chemically, the surface is a crowded marketplace. The interior of a semiconductor has a clear assignment of bonds; atoms "know" how many neighbors they should have. At the surface, that count changes. An atom might be missing a neighbor it would have had inside the lattice, and the leftover bonding capacity can manifest as an electronic state. Ligands—small molecules bound to the surface—are supposed to satisfy these dangling bonds, but they come with their own complexity. They can bind strongly or weakly, cover the surface densely or sparsely, and respond to light, oxygen, water, and heat. In solution, ligands are in equilibrium: they can detach and reattach, allowing the surface to breathe chemically. In a film, ligands may be partially displaced by the host material or by processing steps. This dynamism is part of the reason a dot that performs beautifully in a vial can underperform when turned into a device layer.

The trap story is often told as if traps simply "exist" or "do not exist," but real samples occupy a continuum. Some dots have a small number of shallow traps that release carriers quickly back to the band edge; others have deep traps that hold carriers long enough to encourage nonradiative decay. A particularly frustrating regime is when trapping is intermittent, turning an emitter into a fickle source whose brightness wanders over time. Even without addressing the single-particle flickering associated with blinking, there can be slow spectral diffusion: slight shifts in emission energy as the local surface charge environment changes. Those shifts may average out in ensembles but can broaden linewidths and complicate precise color control.

One reason surfaces are so treacherous is that they offer multiple locations for trouble. It is tempting to imagine the dot as a pristine core with a single outer surface. Many practical nanocrystals are more layered. A core may be coated with a shell of a second semiconductor to improve stability and brightness, and that introduces a buried core/shell interface. The interface can be nearly perfect if lattice constants match and growth is controlled, or it can be riddled with dislocations and strain if the mismatch is large. Those interface defects can act as traps, sometimes more insidious than outer-surface traps because they sit closer to the region where the exciton wavefunction lives.

The shell itself can harbor problems. A shell is meant to be protective, but it is still a crystal grown under finite temperature and chemical constraints. It can contain vacancies, interstitials, alloy fluctuations, or grain boundaries. A thick shell helps in many cases because it moves the exciton wavefunction away from the outer surface and adds a barrier against environmental attack, but thickness is not a magic eraser. Shell defects can persist, and thick shells can even accumulate their own defects if the growth conditions are not well controlled. Not only the outer surface matters: buried core/shell interfaces and shell defects can create traps; thick shells help but do not guarantee high optical performance if interface and shell defects persist.

This multi-layer defect geography is one reason progress in nanocrystal brightness has often come in leaps rather than smooth increments. A small tweak to a synthesis can change not only the average size but also the surface termination, the stoichiometry, and the prevalence of certain defect motifs. The optical outcome may look disproportionate: a modest chemical change yields a startling improvement in photoluminescence. Early practitioners learned this lesson in an especially visceral way. A flask that had been clear and glowing could, after a seemingly innocuous purification step, become dull.

The color peak might stay nearly the same—size had not changed much—but the intensity could collapse because the surface had been disturbed. More than one chemist has described the experience as a kind of betrayal: the synthesis "worked," yet the particles refused to shine.

A concrete example comes from the practical transition between a lab curiosity and a product. In a display film, nanocrystals are often dispersed in a polymer matrix. The particles must survive not only light but heat, oxygen, and the slow stresses of a device's lifetime. The polymer may contain residual catalysts or peroxides; processing may involve UV curing; the film may be laminated under pressure. Each step is an invitation for surface ligands to be displaced or chemically altered. If that creates new dangling bonds or changes local stoichiometry, trap states appear, and k_{nr} rises. The screen does not fail spectacularly; it simply dims, or its colors drift, or its efficiency falls below a specification that was tight to begin with.

The same vulnerability plays out in imaging, where the environment is not a polymer but a biological fluid. Salts screen charges, proteins stick to surfaces, and pH changes can protonate or deprotonate ligands. A dot that is well passivated in a nonpolar solvent can become a mess in water unless its surface chemistry is engineered to be robust. If a trap opens up in the wrong place, the particle may still absorb strongly but return fewer photons, which in a microscope translates directly to longer exposures, higher illumination intensities, and a narrower margin between signal and background.

The early-time nature of trapping also clarifies why "fast relaxation" does not guarantee strong emission. In the ideal narrative, an exciton cools quickly to the band edge and then waits nanoseconds to emit. If a surface trap can capture a carrier in a few picoseconds, it competes not with the nanosecond radiative process but with the cooling and relaxation steps themselves. This is a different kind of fragility. It means that the surface is not only a destination for carriers that have wandered; it can be an intercept point that short-circuits the entire photophysical sequence.

Seen this way, a nanocrystal's brightness is the outcome of a race with multiple legs: excitation, relaxation, trapping, and recombination. The surface is positioned to interfere with each leg because it is close to everything. The dot is small enough that wavefunctions can touch the surface even when carriers are notionally "in the core," and small enough that chemical changes on the exterior can reshape the energy landscape inside. The extraordinary sensitivity of these materials—their ability to have properties tuned by a fraction of a nanometre—comes with a mirror sensitivity to defects and disorder that are also measured in fractions of a nanometre.

Yet it would be a mistake to treat surfaces as merely a nuisance. They are also a handle. The same chemical accessibility that allows traps to form allows them to be repaired, suppressed, or strategically isolated. The surface is where chemists can intervene most directly, swapping ligands, adjusting stoichiometry, growing shells, and controlling interfaces. The glow that looks effortless in a finished product is usually the result of painstaking control over this boundary zone: a deliberate effort to make the surface behave as if it were not there, even though it constitutes so much of the particle in the first place.

A nanocrystal that shines well is, in a sense, one whose surface has been persuaded to remain quiet—to stop offering alternative states, to stop grabbing carriers at the worst possible moment, and to stop turning absorbed energy into heat. When that persuasion fails, the dot does not just get a little worse. It can go from a precise, size-tuned emitter to a dissipative speck whose most reliable output is warmth. That is the double-edged nature of smallness: the same geometry that gives exquisite control over color also places the fate of every excitation within a few nanometres of the outside world.

3.4 Sealing the Leaks: How Chemists Teach a Dot to Behave

Once you start looking for where light goes missing, the culprit stops being mysterious. The losses are not sprinkled evenly throughout a nanocrystal as if the entire material were a little bit flawed. They cluster at particular addresses: dangling bonds, undercoordinated atoms, rough interfaces, and chemically restless patches where the outside world can grab hold. The reason passivation feels so transformative is that it targets those addresses directly. It is not a decorative coating; it is a way of changing the available escape routes for an excited state.

The basic bargain of passivation is simple. A dot owes its tunable color to confinement, but confinement also pushes the exciton wavefunction up against imperfections. If the surface offers trap states, the excited carriers can be diverted into nonradiative loss channels long before a photon is emitted. Ligands and shells act as leak sealants by passivating surface dangling bonds and physically and electronically separating carriers from surface traps. That phrase hides two distinct strategies. One is chemical: satisfy the surface bonds so that trap states do not exist as energetically accessible destinations. The other is spatial: even if some defects remain, move the exciton away from them, or raise barriers that make it harder for carriers to reach them.

Ligands are the first line of defense, and they are easy to underestimate because they are small. In colloidal synthesis the dot is born into a soup of organic molecules that coordinate to its surface. Those molecules keep particles from sticking together, but they also determine whether the surface atoms are electronically "complete." A well-chosen ligand can donate electron density or accept it, match the surface atom's preferences, and stabilize a stoichiometry that avoids mid-gap states. A poorly chosen ligand can leave patchy coverage, bind weakly so it falls off during purification, or introduce new reactive sites that invite oxidation.

That last point matters because the surface is not static. A nanocrystal in solution is not a jewel in a display case; it is more like a small city with constant traffic at its borders. Ligands exchange. Solvent molecules compete for binding sites. Trace water and oxygen can react. Heat and light can accelerate all of it. From the outside, the sample looks unchanged—still the same color, still the same concentration—but on the scale that matters, a few missing ligands are enough to open a trap. When brightness drops after a processing step, it is often because that step altered the traffic pattern at the surface.

Chemists developed a repertoire of ways to make ligand passivation more robust. Some approaches rely on choosing ligands that bind more tightly or cover the surface more completely. Others use chemical treatments that repair stoichiometric imbalances or neutralize undercoordinated sites. The details vary with material—what counts as a "good" ligand for one semiconductor may be useless for another—but the guiding idea is consistent: keep the surface from advertising electronic states in the gap.

Ligands alone, however, are rarely enough for the most demanding applications. They are organic molecules operating in environments that may be hot, bright, oxidative, or mechanically stressful. For display films, the particles must survive years of intense illumination and elevated temperature. For biological imaging, they must remain stable in water, salt, and protein-rich fluids. In those settings, the more powerful intervention is architectural: grow an inorganic shell.

A core/shell nanocrystal is an engineered object in the literal sense.

Instead of a single semiconductor particle, you have a core that sets the emission energy surrounded by a second semiconductor with a wider band gap. The shell is not merely a physical barrier, though that helps. It reshapes the electronic landscape so that the carriers responsible for emission are energetically discouraged from reaching the outer surface. In the simplest picture, the shell is a higher-energy material that acts like insulation: it keeps the exciton's wavefunction concentrated in the core, away from the chemical chaos outside.

This is the same logic as putting a fragile instrument in a protective case, with one important twist. A case does not change the instrument's internal resonances; a shell does change the carrier wavefunctions. It can alter radiative rates, modify how sensitive the emission is to external charges, and change how strongly the exciton "feels" the environment. That is why shell material, thickness, and the sharpness of the interface matter so much. Passivation is not just covering the surface; it is reshaping the probability landscape of where an excited carrier can go.

The first wave of core/shell designs became famous for a reason: they worked. CdSe cores capped with a wider-band-gap shell, often ZnS, produced much brighter, more stable emitters than bare cores. The improvement was so striking that it helped turn nanocrystals from a lab curiosity into a realistic technology platform. The intuition was straightforward: ZnS has a larger band gap than CdSe, so it should confine carriers to the CdSe core and protect it from surface traps. For many purposes, this was exactly the right direction.

Yet the story never stays that tidy at the nanoscale. Epitaxial shell quality is limited by lattice mismatch and strain that can introduce defects. CdSe and ZnS do not have identical lattice constants. When you try to grow one crystal on another with a mismatch, strain builds. In a thick film of macroscopic materials, mismatch is handled by dislocations and other defects; at the nanoscale, strain can be partially accommodated elastically, but only up to a point. Push too far and the shell may crack, roughen, or develop defects that create traps of their own. A shell meant to seal leaks can, if poorly matched, introduce new holes in the plumbing.

This is why the comparison between CdSe/ZnS and CdSe/CdS became more than a materials footnote. CdS is closer to CdSe in lattice constant than ZnS is, which can make it easier to grow thicker, higher-quality shells with fewer strain-induced defects. That does not mean CdSe/CdS is automatically better; CdS also changes the band alignment and can allow the exciton wavefunction to spread into the shell, depending on thickness and composition. But the tradeoff is instructive. A "better barrier" in band-gap terms (ZnS) can be a "worse barrier" in defect terms if the interface is riddled with strain. A "softer barrier" (CdS) may allow more wavefunction leakage into the shell but offer a cleaner crystal landscape with fewer traps. In nanocrystals, the cleanest interface often wins.

That interface—the buried boundary between core and shell—deserves its own attention. It is tempting to think that if the outer surface is passivated, the job is done. But not only the outer surface matters: buried core/shell interfaces and shell defects can create traps. Carriers do not need to reach the outside world to be lost; they can be captured at an internal boundary if that boundary contains dangling bonds, compositional roughness, or dislocations. And because the interface sits close to the exciton's natural habitat, even a small density of defects there can have an outsized effect.

As core/shell design matured, chemists began to treat the interface as something to be engineered rather than tolerated. Shell growth procedures were refined to reduce roughness, avoid sudden compositional jumps, and minimize the formation of defect motifs. Alloyed or graded interfaces—where composition changes gradually rather than abruptly—can reduce strain and smooth the electronic potential.

These approaches are hard to perfect because they depend on kinetics: how precursors decompose, how atoms migrate on the surface, how growth proceeds from facet to facet. The same nominal recipe can produce different results depending on subtle details of temperature, precursor purity, or ligand environment. In other words, architecture matters beyond thickness: shell growth complexity, morphology, and defect density can determine whether thick-shell dots actually achieve suppressed blinking and high stability.

This is where a historical anecdote captures the spirit of the field. When "giant" core/shell nanocrystals—particles with unusually thick shells—began to attract attention, the promise was almost audacious: single particles that would not blink, that would emit steadily instead of flickering unpredictably. Blinking had been one of the most maddening behaviors revealed by single-particle spectroscopy: an emitter could be bright for seconds, then dark for seconds, as if it were deciding whether to cooperate. Thick-shell architectures offered a way to change the rules. By isolating the excitonic wavefunction from surface effects and by reducing access to traps, thick-shell designs could strongly suppress blinking and photobleaching. The threshold-like nature of the improvement was part of the excitement: add enough shell, and the statistics of emission could change qualitatively.

The deeper lesson is not simply "more shell is better." Thick shells are hard to grow without defects, and the shell material and strain budget matter. A giant shell with imperfections can still harbor traps; a thick but poorly matched shell can accumulate strain that generates new nonradiative pathways. The most successful designs treat thickness as one tool among several, combined with careful material choice and growth control so that the added volume genuinely isolates the emissive region rather than surrounding it with a defective cage.

There is a further benefit to this kind of architectural passivation, one that becomes crucial when emitters are driven hard: chemical and structural strategies can also reduce Auger recombination, a multicarrier loss process that becomes severe when a nanocrystal is charged or contains more than one exciton. Auger recombination is a particularly punishing way to lose light because it can be much faster than radiative decay; the energy that could have produced a photon is instead handed to another carrier, which then dissipates it as heat. If a dot is used in a regime where charging occurs or where multiple excitons are created, suppressing Auger pathways can stabilize brightness under continuous illumination and keep emission efficient even in charged states.

Core/shell design can influence Auger processes because Auger rates depend sensitively on confinement and on how wavefunctions overlap. "Softening" the confinement—allowing wavefunctions to spread out into a larger volume—can reduce the strength of Coulomb interactions that drive Auger recombination, lowering the probability that recombination energy will be dumped into a bystander carrier. Thick shells can provide that extra volume. Material choice and band alignment matter as well: if one type of carrier is preferentially delocalized into the shell, the overlap that enables fast Auger transfer can be reduced. This is one reason thick-shell architectures have been so attractive for applications that demand high brightness under intense pumping, from microscopy to emerging light sources.

It can sound as if chemists are merely adding layers until problems go away. In practice, the work feels more like negotiation with a stubborn object. Every improvement introduces new constraints. A shell that improves optical stability might worsen processability by making particles larger and harder to disperse. A ligand that binds strongly might impede charge transport if the particles are used in an electronic device. A graded interface that reduces strain might be more complex to synthesize reproducibly. And the environment always gets a vote: a dot that behaves in a nonpolar solvent may behave differently in a polymer film or in water, because ligands reorganize and surfaces encounter different chemical challenges.

The most revealing way to think about successful passivation is not as "adding protection" but as "editing the list of allowed outcomes." In the rate language used earlier, you are trying to reduce k_{nr} without damaging k_{r}, and in some designs you may also be reshaping rates associated with trapping and Auger loss. Ligands and shells work when they remove energetic stepping stones that lead into traps, and when they reduce the probability that carriers even approach regions where defects lurk. The result is not just higher brightness on day one, but a kind of behavioral training: an emitter that, under real-world conditions, continues to do the one thing you want it to do—return energy as light—rather than exploring all the other options available to matter.

A well-passivated nanocrystal is therefore an engineered compromise between chemistry and quantum mechanics. It accepts that the surface will always exist, that the environment will always prod it, and that defects will never be eliminated entirely. The goal is to make the emissive state the easiest, most natural outcome—to turn radiative recombination into the path of least resistance, so that when a dot is jolted into an excited state, it settles down by paying back a photon instead of quietly bleeding energy into the surrounding world.

3.5 Blink, Burst, Go Dark: When One Dot Refuses to Be Predictable

A film of nanocrystals can look wonderfully well behaved. Illuminate it, and it emits a steady wash of color that seems as dependable as a lamp. That steadiness is partly an illusion created by crowds. When many emitters contribute at once, their individual quirks average out. The surprise came when researchers learned to watch a single particle at a time. Under a microscope, with the background suppressed and the photon stream counted, one nanocrystal could act like a temperamental lighthouse: bright for a while, then suddenly dark, then bright again, without warning and without any obvious change in the illumination.

This intermittency became known as blinking. It was not a minor technical wrinkle. It was a conceptual jolt, because the ensemble picture suggested that each particle was a tiny, repeatable light source. Blinking showed that an individual emitter can jump between distinct modes of behavior, even when nothing macroscopic changes. The particle is not merely producing photons with some average probability; it is switching between internal conditions with very different odds of giving light back.

The basic observation is easy to describe. Record the emission intensity from a single dot as a function of time under constant excitation. The trace is not a noisy line around a steady value; it often contains plateaus. For seconds or milliseconds the intensity sits high, then it drops to near zero, then it returns. Early single-particle experiments made this impossible to ignore because the switching was so stark. It felt almost alive: as if the particle had moods.

The physics behind blinking is less theatrical but more instructive. Authoritative reviews converge on a central idea: blinking is commonly attributed to switching between emissive neutral exciton states and non-emissive or weakly emissive states caused by charging (Auger quenching) and/or trapping pathways. Those two mechanisms can cooperate and compete. Sometimes a carrier is captured by a trap, leaving the particle charged; sometimes charging arises by other photoinduced processes and then makes trapping more likely.

Sometimes a trap intercepts an excited carrier before it ever reaches the band-edge emissive state. In all cases, the common feature is that the dot's internal ledger of rates suddenly changes. The radiative route is not physically removed, but it is outcompeted.

Charging is the cleanest route to a "dark" state to explain at a dinner table, because the consequence can be dramatic. A neutral dot under modest excitation typically holds at most one exciton at a time. If the dot becomes charged—say it has an extra electron or an extra hole—then absorbing another photon can create a charged exciton (often called a trion). Now the system contains three carriers instead of two. When recombination occurs, the energy that would have left as a photon has another place to go: it can be transferred to the extra carrier. That transfer is Auger recombination, named after Pierre Auger, who uncovered related processes in atomic physics in the 1920s. In an Auger event, recombination energy is converted into kinetic energy of a bystander carrier rather than light. The bystander then rapidly dumps that energy as heat to the lattice and environment. The result is a fast nonradiative channel that can overwhelm photon emission.

Auger recombination is not a rare side process in nanocrystals; it can be brutally efficient because confinement increases the overlap between carriers. Put three carriers in a small volume and their interactions become strong. The rate can become so high that the radiative lifetime—normally leisurely on nanosecond scales—no longer matters. The dot can be "on" when neutral and "off" when charged, not because it cannot emit at all when charged, but because Auger recombination becomes a shortcut that wins the race.

This picture also explains why blinking is often sensitive to excitation intensity. Stronger pumping makes it more likely to create multicarrier states and more likely to drive photocharging pathways. Even when a particle is mostly neutral, intense excitation can generate biexcitons: two excitons at once. A biexciton contains four carriers. That is an invitation for Auger recombination, and it often answers. This is why a dot can be bright under gentle illumination and yet perform poorly when pushed hard, exactly the regime that high-brightness applications demand.

Trapping-based explanations emphasize a different vulnerability: the surface and interface defects discussed earlier do not merely steal energy in a smooth, averaged way. They can also open and close dynamically. A trap may be available in one moment and passivated in the next, depending on local rearrangements of ligands, adsorption of molecules from the environment, or the slow wandering of charges nearby. In that view, the dot's dark state is not always a charged state dominated by Auger loss; it may be a period in which excited carriers are intercepted by traps before they can emit. The nuance matters because different mechanisms predict different experimental signatures. Some dark states correlate strongly with shortened lifetimes (consistent with fast nonradiative channels); others show more complex behavior, including "gray" states where the dot continues to emit but at reduced intensity or with altered lifetime.

That "grayness" is important, because blinking is not always a clean binary. Many single-particle traces show intermediate levels, suggesting partially emissive charged excitons or multiple competing nonradiative routes. The temptation is to force every particle into a simple two-state model—on and off with fixed rates—but nature resists. Blinking kinetics frequently show complex statistics, often with power-law-like distributions of on and off times. Instead of having a characteristic off-time, the particle can have off-periods spanning a broad range, from microseconds to minutes, with no single timescale dominating. Such behavior implies broad distributions of trap depths and return times, or fluctuating environments that continuously reshape the landscape. A single "defect level" is not enough;

the surface is a disorderly network of possibilities.

This is where the ensemble view can be dangerously comforting. If you average over millions of particles, each with its own traps and its own blinking rhythm, the output looks steady and well behaved. The device engineer may never notice intermittency at the single-particle level, yet the same underlying physics can still matter. Blinking is a symptom of switching among emissive and non-emissive pathways. Even if you never watch one dot in isolation, those switches can contribute to noise, to long-term instability, and to efficiency loss under real operating conditions.

A historical anecdote from the rise of single-molecule spectroscopy captures why the field took blinking seriously. Long before nanocrystals became common in consumer devices, chemists learned to detect fluorescence from individual dye molecules. One of the marvels of that work was that you could "see" a single molecule by its photons. But those molecules often photobleached: after some time they would permanently go dark because a chemical reaction destroyed the fluorophore. Nanocrystals seemed, at first, like a solution—more robust, less prone to irreversible chemistry. Blinking complicated the optimism. The emitters did not always die, but they did disappear and reappear, which in a tracking experiment can be almost as damaging as photobleaching. If you are trying to follow a labeled protein moving through a cell, a label that vanishes for seconds may as well have detached.

The drive to suppress blinking became a design challenge with real emotional stakes: make the label reliable enough that a biologist can trust what they see, or make the emitter stable enough that a display does not show subtle time-dependent artifacts. The solutions led back to the same theme as passivation, but with sharper demands. It is not enough to raise average brightness; the goal is to stabilize the *state* of the emitter.

Thick-shell "giant" dots became famous partly because they changed the single-particle experience. By isolating the excitonic wavefunction from surface effects and reducing access to traps, thick-shell architectures can strongly suppress blinking and photobleaching. The mechanism is not a single magic step. A thick shell can reduce the probability that a carrier reaches the outer surface where trapping and charging events are likely. It can also soften confinement and alter wavefunction overlap, which can reduce Auger rates. In other words, the shell can both prevent the creation of problematic states and lessen the penalty when they occur.

Even here, nature keeps a footnote. Thick-shell dots can appear "nonblinking" in intensity while still showing lifetime fluctuations—sometimes called lifetime blinking—implying hidden switching among emissive states even when intensity is stable. The emitter may continue to look bright, yet its radiative and nonradiative rates can vary in ways that only reveal themselves when you measure not just how many photons arrive, but how quickly they arrive after each excitation. For some applications, intensity stability is enough. For others—precise time-resolved imaging, quantum optics experiments, or metrology—lifetime fluctuations are a reminder that "fixed brightness" is not the same as "fixed physics."

Auger recombination, meanwhile, is not only a villain in blinking. It is the central obstacle to high-excitation performance more broadly. Any technology that demands that a dot emit lots of photons quickly—high-brightness LEDs, potential nanocrystal lasers, fast scintillators—runs into the same constraint: once multicarrier states are populated, Auger loss can dominate. This is a cruel trade. To get more light out, you pump harder and create more excitons; by doing so you increase the chance that the dot routes energy into Auger heating instead of photons.

The most productive way to think about Auger recombination is as an internal energy transfer tax that becomes expensive when the particle is crowded with carriers. In a neutral single-exciton regime, the dot can afford to wait nanoseconds and emit. In a charged or biexcitonic regime, the dot is offered a much faster exit that does not involve light. The "tax rate" depends on architecture. Softening confinement, spreading wavefunctions over a larger volume, and designing interfaces that reduce carrier overlap can lower Auger rates. That is why shell design is not only a protective measure against surface traps but also a strategy for taming multicarrier physics.

There is also a productive ambiguity in the blinking story that researchers continue to navigate. Debate and nuance remain: some work argues photodarkening and off-states may involve trap-mediated cycles rather than being governed purely by Auger recombination, suggesting multiple mechanisms may dominate in different materials and conditions. The implication is practical. If blinking in a particular system is primarily driven by charging and Auger quenching, then strategies that suppress Auger processes and prevent charging should help. If blinking is primarily driven by trapping and slow detrapping dynamics, then the focus shifts to eliminating traps and stabilizing the surface environment. Often both are needed, and the balance can change with excitation wavelength, intensity, temperature, and the chemical surroundings.

All of this might sound like a technical coda, but it changes how one should interpret "brightness" and "efficiency" claims. An ensemble measurement can report a high quantum yield while hiding a minority population of particles that blink severely or charge easily. In some contexts that minority hardly matters; in others, it dominates the failure mode. A single emitter used as a quantum light source, for example, cannot average over its own bad moments. A particle used as a biological tag cannot rely on its neighbors to fill in the darkness.

The deeper message is that a nanocrystal is not just a static object with a fixed band gap. It is a small system whose charge state can fluctuate, whose surface can reorganize, and whose internal interactions change depending on how hard it is driven. At the ensemble level, the glow looks like a smooth property of a material. At the single-particle level, it becomes a conversation between chance and design: every absorbed photon is an opportunity, and every nonradiative pathway is a temptation. When a dot blinks, it is not misbehaving out of spite; it is revealing that, in a world of traps, charging, and Auger processes, "glow" is a precarious outcome that must be engineered into a habit.

Chapter 4

From Serendipity to Stockholm: The Road to the 2023 Nobel

The chapter opens with an experimental surprise: tiny crystals refusing to behave like ordinary bulk matter, changing color simply because their size changed. As the story unfolds, readers see how a curious optical anomaly hardened into a new scientific intuition, then into a manufacturing platform—until Stockholm recognized not a single eureka moment, but a decades-long chain of insight, technique, and control.

4.1 When Glass Began to Lie About Color

For most of modern science, color has been a comfortingly reliable witness. Copper salts turn flames green; cobalt makes glass blue; gold in bulk is stubbornly gold. If a material changed color, there was usually a straightforward explanation: different atoms, different impurities, a different crystal structure, perhaps a change in oxidation state. Color, in other words, seemed to belong to chemistry in a stable, almost bookkeeping sense. You could point to a sample and say, with confidence, *this* is what it is, therefore *this* is the color it must have.

The early years of nanocrystal research contained a small heresy: the atoms could stay the same, and the color could still shift.

It is hard to appreciate how disorienting this was without remembering what "normal" solids look like to a physicist or chemist trained on bulk materials. In a macroscopic crystal, the relevant energies form bands: broad, quasi-continuous ranges of allowed energies for the material's mobile charges. The gap between the filled valence band and the empty conduction band—the band gap—is the decisive number for what colors the solid can absorb, and therefore (indirectly) what colors it can emit or transmit. In the bulk, that number is a property of composition and crystal structure. Silicon has its band gap; cadmium selenide has its band gap; they are not supposed to slide around because you happened to make a smaller piece of the same thing.

The first hints that this tidy picture could fail arrived in a setting that seems, at first glance, almost too ordinary to host a revolution: glass.

Glass is a strange solid—rigid, but not crystalline. It is also a venerable laboratory for surprises because it can be doped, heated, and coaxed into doing things crystals resist. For decades, glassmakers and materials scientists had known that tiny particles embedded inside glass can create vivid colors. Medieval stained-glass windows owe their reds and yellows, in part, to minute metal clusters and nanoparticles. But those older colors were not usually read as quantum mechanics made visible; they were treated as a complicated, but ultimately classical, interplay of light with small inclusions.

In the late 1970s and early 1980s, Alexei Ekimov and collaborators in the Soviet Union were studying glasses containing semiconducting compounds. The experiments, in one sense, were straightforward materials science: prepare a glass with certain ingredients, heat-treat it, and see what optical features appear. Heat treatment matters because it allows atoms to move. Under the right conditions, the doped components in the glass can separate out into tiny crystalline regions—nanometer-scale islands of a semiconductor trapped in a rigid, transparent host.

That last detail is crucial. A glass matrix can act like a tiny nursery: it permits particles to form and grow but prevents them from quickly clumping into larger chunks. In a liquid, particles can drift together and fuse; in a glass, they are pinned in place. The result is a population of nanocrystals that can remain stable long enough to measure carefully.

Ekimov and colleagues observed that the optical absorption of these glass-embedded semiconductor crystallites changed with the heat-treatment conditions that controlled their size. What appeared in the spectra was not just a subtle variation in intensity. Peaks and edges shifted in a way that implied a change in the fundamental energy needed to excite the material. The sample's *color*—in the broad sense of which wavelengths it absorbed—depended on how large the crystallites had become inside the glass.

This was the point at which glass began to "lie" about color, because the everyday intuition says that glass acquires its color from what is dissolved in it, not from an invisible architectural detail like the size of particles suspended within. To be sure, glassmakers were familiar with the idea that particles can tint glass; what was new was the implication that *the same semiconductor*, with the same composition, could be tuned through a range of optical behaviors merely by changing the particle's dimensions.

In bulk semiconductors, the band gap is often introduced as a fixed rung in an energy ladder. But the experimental reality that Ekimov's group brought into focus was that, at sufficiently small sizes, the ladder's spacing itself changes. The nanocrystal behaves less like a miniature version of the bulk and more like a system with its own internal rules, with allowed energies that can shift as the physical boundaries move inward. This was not just a curiosity about color; it was a concrete, measurable sign that the familiar band picture had to be supplemented with something else when matter was confined.

A helpful way to feel the strangeness is to imagine a jeweler who swears a gemstone's hue is determined by its chemical identity, and then watches that hue change in a batch where the chemistry is held constant. The first reaction is to suspect contamination. The second is to suspect measurement error. Only later does the mind accept the unsettling alternative: the property you thought was intrinsic can be tuned by geometry.

Behind the spectral shifts was a length scale that had been known in semiconductor physics but had not often been made central in the lab's day-to-day thinking: the typical size of a bound electron-hole pair, an exciton. In many semiconductors, when light promotes an electron to a higher-energy state, it leaves behind a positively charged "hole" in the valence band. The two attract each other, forming a correlated pair that can have a characteristic spatial extent. When the physical dimensions of the crystal approach that extent, the pair's motion is constrained; the energy of the excitation shifts. As mentioned earlier, this is the heart of the confinement effect, but the early glass experiments gave it an unusually visceral form. You could not see the nanocrystals directly in the glass with the naked eye, yet their presence and size announced themselves through changes in color.

What made this particularly striking is that the experiment did not begin as an attempt to build a new kind of light emitter. It began as materials science: heat, cool, measure. Serendipity in the strongest sense is not a lightning bolt from nowhere, but a moment where an experiment built for one purpose answers a different question more loudly than the one you asked.

There was also a sociological twist embedded in the glass itself. Research cultures can be separated by the materials they handle. People who work on crystalline semiconductors often live in a world of wafers, epitaxy, and carefully controlled crystalline order. People who work on glass inhabit a different landscape: amorphous hosts, phase separation, heat treatment protocols, and the practicalities of optical transparency. A phenomenon that appears in glass can, for a time, feel like a "glass phenomenon"—a peculiarity of that messy, industrially tinged material—rather than a general principle about nanoscale semiconductors. It took a second strand of evidence, arriving from a different community, to make the point unavoidable.

That second strand came from the chemistry side, with Louis Brus and colleagues in the early 1980s. Instead of nanocrystals imprisoned in glass, Brus worked with small semiconductor crystallites in solution. This was a different experimental universe. A solution-phase sample could be probed not only by its optical spectra but also by chemical behavior—redox potentials, surface reactions, and the effects of the surrounding solvent and ligands. The question was no longer just "what light does it absorb?" but "how do its energies line up against chemical reference points?" In 1983 and 1984, Brus developed models linking the size of these crystallites to their excitation energies and related electronic properties.

The move from glass to solution mattered because it brought a second kind of credibility. A result confined to a solid matrix could always be doubted: perhaps strain in the glass altered the semiconductor; perhaps the particles were defective; perhaps the host material played an essential role. A result that survived in a chemically active environment, and that could be connected to simple theoretical scaling, looked less like an artifact and more like a general nanoscale rule.

Yet the solution experiments had their own complications, and they reveal another reason early progress was slow and sometimes contentious. Making nanocrystals in solution in the early 1980s did not reliably yield uniform particles. A sample might contain a spread of sizes: some tiny, some larger, some aggregated. When that happens, the spectrum becomes a kind of census report written in overlapping ink. Peaks broaden. Edges smear. The signal of size-dependent behavior is present, but it must be teased out from disorder.

This is where glass played an unsung role. A glass matrix tends to produce relatively stable, isolated crystallites whose growth can be controlled by temperature and time. Even if the size distribution was not razor-sharp, the particles were physically separated and protected from the quick catastrophes of colloids—clumping, ripening, oxidation. Glass did not merely host the effect; it made it measurable at a time when synthetic control in solution was still primitive.

To a reader used to thinking of glass as inert, it is worth lingering on the physical picture. During heat treatment, small regions enriched in the semiconductor components can form and crystallize. The glass around them constrains how they grow: it is like baking tiny crystals into a transparent loaf. Because the host is transparent, you can send light through and read the nanocrystals' absorption features. Because the host is rigid, you can store the sample, remeasure it, and compare it across heating conditions. The glass becomes a laboratory notebook written in color.

The deeper conceptual shift, however, was not about experimental convenience. It was about what counted as a "property." If the band gap could be tuned by changing size, then size was not just a geometric descriptor. It became a control knob as real as composition. That was a new category of thinking for many chemists and materials scientists: not simply "what is it made of?" but "how big is it?" could determine fundamental energetic behavior.

At an intuitive level, the reason is easier to accept when you picture motion rather than abstract energy. In a macroscopic crystal, the excited charge carriers can roam over enormous distances compared with the crystal's unit cell. In a very small crystallite, that roaming is curtailed. When motion is confined, the allowed patterns of motion change, and energy shifts follow. This is familiar from other corners of physics, but it had not been a daily tool for people who built materials out of beakers and furnaces. The shock was less that quantum mechanics could do this—quantum mechanics can do almost anything—than that it could do it in a way that showed up as a change in visible color in a piece of glass.

There is a concrete way to see why "color" is such a powerful messenger here. Optical photons carry energies comparable to electronic excitations in semiconductors. A small shift in the excitation energy corresponds to a noticeable shift in wavelength. That means the nanoscale does not merely perturb a property; it can move it across the range our eyes and instruments care about most. If you are studying magnetism or mechanical strength, a nanoscale effect might hide behind complicated measurement protocols. With optical absorption, the effect can announce itself immediately as a change in the spectrum, and sometimes as a visible tint.

That visibility carried a psychological consequence. A scientist can intellectually accept that energy levels are quantized while still treating bulk matter as essentially continuous in practice. The glass experiments made quantization feel less like a theoretical overlay and more like a tangible lever. The sample's appearance became, in a sense, a readout of nanoscale physics.

The early reports did not instantly convert the wider scientific community. The 1980s were full of claims about "small particles" and "microcrystals," and not all of them held up. Skepticism was rational: if the color changes depend on size, then you must be sure size is actually what you have changed. But the essential seed had been planted. There existed a class of materials—semiconductors, prepared as nanocrystals—whose optical behavior could be tuned without changing composition, by shifting the physical boundary conditions of the crystal itself.

That seed would eventually grow into a technology that now sits behind millions of living-room screens and a growing set of biomedical tools. In the early 1980s, though, it was enough that a piece of doped glass, carefully heat-treated, could force a new thought: maybe color is not always a fingerprint of chemistry. Sometimes it is a ruler for size, with quantum mechanics holding the ruler in place.

4.2 From Curiosity to Credibility

An anomalous spectrum can be a trap. Anyone who has spent time around real instruments knows how easily nature can be impersonated by dust, by a misaligned lamp, by a cuvette with a faint tint, by an impurity that arrived with a reagent and never introduced itself. A new field does not begin when someone sees something strange; it begins when enough people, in enough places, can make the strangeness appear on demand—and can explain why it appears where it does.

The early color shifts in nanocrystals embedded in glass had the glamour of a paradox: keep the chemistry the same and still change the optical response. But the scientific community's reflex was to ask a series of hard, almost mundane questions. Were the tiny crystals really the same material each time? Were they uniform, or was the signal dominated by a small fraction of outliers? Could the same sample be prepared twice? Could a different laboratory, with different furnaces and different glass recipes, see the same pattern? Most importantly, did the color shift obey a rule that could be checked, stretched, and eventually used?

The slow transformation from curiosity to credibility happened through a particular kind of evidence: repeated spectra that lined up with particle size. Not a single dramatic curve, but a stack of them, each one nudged a little, like a melody transposed into higher keys. When that stack appeared again and again—across materials, across preparation methods, across research groups—it stopped looking like a trick of glass or a quirk of a single setup. It began to look like a principle.

A useful starting point is to separate two ideas that often blur together in casual conversation: absorption and emission. Absorption is about what wavelengths a material *removes* from incoming light; emission is about what wavelengths it *produces* after being excited. In semiconductors, absorption is tied to the threshold energy needed to lift charges across the band gap. Emission, in contrast, is a story about what happens after excitation: how the system sheds energy and returns toward its ground state.

In the laboratory, absorption spectra are often easier to trust early on. They do not require a bright luminescent material; they require only that the sample modify a beam of light in a consistent way. Emission—photoluminescence, the glow after shining light on a sample—demands additional virtues. The excited state must survive long enough to radiate; the surface must not offer too many non-radiative escape routes; the material must not self-destruct under illumination. Early nanocrystals did luminesce, but not always cleanly or strongly, and that unevenness fed skepticism: if a phenomenon cannot reliably produce light, perhaps the claimed "size-tunable" behavior is more fragile than advertised.

What settled the argument was not a single perfect nanocrystal, but a pattern: absorption features shifting systematically with size, and photoluminescence following along when the material quality allowed it. With each additional dataset, the field gained a new kind of confidence, the confidence that comes from a control knob that turns smoothly rather than by jerks.

Louis Brus's work in the early 1980s supplied something that spectra alone could not: a model-driven reason to expect a lawful dependence on size. In 1983 and 1984, Brus connected the size of small semiconductor crystallites to their electronic energies in a way that made concrete predictions. The details can be mathematical, but the conceptual message is easy to state at a dinner table: shrink the crystal and the energy of the lowest excitation rises, partly because confinement raises the kinetic energy, and partly because the mutual attraction of the excited charges offsets that rise. Put differently, there are competing contributions, and their balance depends on the particle radius. That picture offered more than an explanation-after—the-fact; it gave researchers a sketch of what to look for, and what kind of scaling with size might plausibly emerge.

This theoretical framing did something subtle but powerful. It recast the color shift as an energetic shift that would show up in places other than optics. Brus explicitly linked confinement physics to chemical observables such as redox potentials. Redox potentials are, in essence, measures of how eager a system is to gain or lose electrons—an energetic preference dressed in electrochemical clothing. If a nanoparticle's key electronic energies move with size, then its willingness to participate in electron-transfer reactions should move with size too. The phenomenon was no longer merely about pretty colors; it was about tunable energetics. That made the small-crystallite regime chemically consequential, not just optically intriguing.

There is a practical reason this mattered for credibility. Many experimental artifacts can shift an optical spectrum: a change in solvent, a change in scattering, a change in aggregation. It is harder for an artifact to mimic a coherent set of linked shifts across multiple kinds of measurement. When optical data and electrochemical reasoning pointed in the same direction—size pushing key energies in predictable ways—the interpretation tightened. A nanocrystal that absorbs bluer light and also sits at a different energetic position for electron transfer is not just a colored speck; it is a small system with reorganized energy structure.

Still, the early years were messy. Laboratories making colloidal semiconductor crystallites in solution often ended up with mixtures: a distribution of sizes, plus surfaces that were not well controlled, plus occasional chemical byproducts that might themselves absorb light. In that context, reproducibility was not a polite ideal; it was the central battleground. A spectrum with a broad absorption onset could be read as a genuine ensemble of small particles or as a sign that the sample was ill-defined. Researchers had to learn how to tell these possibilities apart.

One of the most persuasive habits that developed in the field was simply to make *series* rather than *singles*. Instead of producing one sample and describing it as "small crystallites," groups began preparing multiple batches under systematically varied conditions—longer heat treatment, different precursor concentrations, different temperatures—and then measuring spectra across the whole set. If the absorption edge marched steadily in one direction as the conditions were tuned to yield larger particles, that marching was difficult to dismiss. It looked like an underlying rule expressing itself through experimental control.

A concrete laboratory scene helps capture why this was convincing. Imagine a bench with a row of vials held up to the light. Each vial contains a colloidal dispersion—liquid that looks clear at first glance, but with a faint tint. Under a UV lamp, some vials glow more brightly than others. Early on, that glow could be fickle, and it might fade or change with time. But even when emission was uneven, absorption spectra could be measured for every vial. When those spectra were plotted, the absorption onset and peaks did not wander randomly; they shifted in a consistent direction that tracked independent estimates of particle size. A student could repeat the synthesis the next week and reproduce the same ordering. A collaborator could do it in another building with a different spectrometer and still see the same progression. That is what a field begins to lean on: not a single dramatic result, but a dependable ladder of outcomes.

The distinction between absorption and emission also taught researchers a new kind of intuition about imperfection. If absorption moved with size but emission behaved erratically, that was not necessarily a refutation of the size effect. It could be a message about the surface. The interior of a nanocrystal might have the right band structure, but its surface—where the periodic crystal lattice abruptly ends—could host defects and dangling bonds that act as energy drains. An excited system can lose energy as heat rather than light if it finds such a drain. That realization changed the way people talked about "good" samples. A sample could be *good enough* to reveal size-dependent absorption yet *not good enough* to provide clean luminescence. This sharpened the community's standards and shifted attention toward chemical strategies that would later become essential for technology.

Skepticism also had a more conceptual form. In bulk solids, many properties are self-averaging. A centimeter-scale crystal can have defects and still behave like the textbook material because the electrons explore a vast interior compared with the defect density. Nanoscale solids are not so forgiving. When a particle is only a few nanometers across, the surface is not a minor correction; it is a large fraction of the atoms. That raises a reasonable doubt: if surfaces dominate, is the observed behavior truly an intrinsic size effect, or is it a surface effect that merely correlates with size? Early work had to untangle this knot, and the answer turned out to be both sobering and enabling. Yes, surfaces matter enormously—but the systematic size dependence in the fundamental excitation energy remained, and it could be separated from surface-induced quirks by careful comparisons across samples and by improving synthesis and passivation.

A further ingredient in the credibility story is that the size effect was not merely monotonic; it had a recognizable shape. Brus's 1984 treatment of the lowest excited state emphasized that confinement contributions scale strongly with size. You do not need the formula to appreciate the implication: if energy shifts scale with a power of the radius, then doubling the radius should not produce a random change; it should produce a change of a predictable order. This is precisely the kind of expectation that helps a field police itself. When a reported shift is wildly too large or too small for the claimed particle size, researchers have a reason to suspect aggregation, mis-sized samples, or mistaken characterization. A model becomes a filter for reliability.

The phrase "size is a control knob" is often used loosely, as if any nanoscale system automatically becomes tunable. In practice, a control knob is only real when turning it gives a repeatable response. In the 1980s, that meant learning how to measure size in the first place. Direct imaging of nanocrystals existed, but it was not always routine, and different techniques could yield different answers depending on what exactly they were measuring. Spectroscopy itself became a kind of measuring stick: if the absorption onset shifts in a consistent way, it can serve as an indirect size indicator. This created a feedback loop. Better size measurements sharpened the spectral correlations; sharper correlations made spectroscopy a more trusted proxy for size; that proxy, in turn, helped optimize syntheses that produced clearer samples.

There is an understated drama in this feedback loop because it is how a community turns an effect into a craft. A newcomer might look at the final result—a clean set of spectra, tidy peaks, a predictable relationship between particle radius and optical features—and imagine that it was always obvious. It was not. It had to be earned by iteration: make a sample, measure it, discover it is broader than hoped, adjust conditions, measure again, compare with a colleague's data, argue about whether the shoulder in the spectrum is real, discover that oxygen exposure changes the result, learn to exclude air, discover that washing the sample removes a background absorber, learn a better purification method, and only then watch the spectra tighten and align.

The field also benefited from a kind of convergence that is often more persuasive than any single method: similar size-dependent shifts appeared whether nanocrystals were formed in glass matrices or prepared in solution. These were very different environments—rigid host versus liquid, relatively isolated particles versus colloidal dispersions—yet the same central trend emerged. That convergence did not make surface effects irrelevant, but it reinforced the idea that the core phenomenon belonged to the nanocrystal itself.

Brus's linkage to redox potentials quietly expanded the stakes. A size-tunable energetic landscape implies that nanocrystals could be engineered not only to emit a chosen color, but to donate or accept electrons with chosen strengths. That suggestion foreshadowed later ambitions in photocatalysis and solar energy conversion, where the question is not merely what light a material absorbs, but whether the resulting excited charges can drive chemistry. Even before those applications matured, the conceptual shift was immediate: small semiconductor crystallites were not just optical curiosities. They were tunable energetic objects, sitting at the intersection of condensed matter physics and chemical reactivity.

By the end of this credibility-building phase, the community possessed something closer to a shared intuition. If a sample's spectral features looked a certain way, experienced researchers could make an educated guess about whether the particles were too polydisperse, whether surfaces were too defective, or whether the size regime was likely in the strong-confinement limit. Those guesses were not always right, but they were grounded in accumulated patterns rather than hope.

A principle, in the scientific sense, is not a slogan. It is a compact expectation that survives contact with new data. In nanocrystal spectroscopy, the expectation was simple enough to say aloud—smaller crystals shift the key optical energies upward—but rich enough to generate a long sequence of tests. Each time an independent group reproduced the trend, each time a synthesis tweak moved the spectrum in the anticipated direction, each time a model connected size to an energetic quantity that could be measured by another route, the effect became less of a novelty and more of a dependable feature of matter at the nanoscale.

At that point, the color change stopped being a puzzle that demanded special pleading. It became a regularity that could be planned around: a way to design a material by choosing a dimension, and to diagnose a synthesis by reading a spectrum. A small crystal in a vial could be treated not as an unruly byproduct of chemistry, but as a system with tunable, checkable energetics—an object that could be made, measured, and trusted.

4.3 The 1993 Recipe That Turned Wonder into Technology

By the early 1990s, the central promise of semiconductor nanocrystals was clear enough to tempt both physicists and chemists: the optical response could be tuned by changing size. The trouble was that most samples behaved like crowds rather than individuals. A beaker might contain millions of particles, but they were not all the same. Some were slightly bigger, some smaller, some misshapen, some stuck together. In spectral terms, that diversity is unforgiving: a sharp feature becomes a blur, and a blur is easy to argue about. If size is the knob, a wide spread of sizes is like trying to tune a radio while three stations play at once.

The field needed something more prosaic than a new idea. It needed a recipe.

In 1993, a paper in *Journal of the American Chemical Society* by Murray, Norris, and Moungi Bawendi delivered a synthesis that many researchers still describe as a turning point. It used hot-injection organometallic chemistry in a hot coordinating solvent to produce nearly monodisperse cadmium chalcogenide nanocrystals. That phrase—*nearly monodisperse*—was the difference between a phenomenon you could admire and a material you could build with. "Nearly" mattered: it meant that most particles in the sample clustered tightly around a target size. The resulting optical spectra tightened, and the relationship between size and optical energy stopped feeling like an anecdote and started behaving like a specification.

The core logic can be said without lab jargon: cause most particles to be born at nearly the same moment, then let them grow together. In the language chemists use, that is a short nucleation burst followed by controlled growth. When nucleation dribbles on for minutes, newborn particles keep joining the population while older ones are already enlarging, and the size distribution inevitably widens. When nucleation is confined to a brief window, the population's "birthday" is synchronized. Growth can then be guided by time and temperature in a more orderly way.

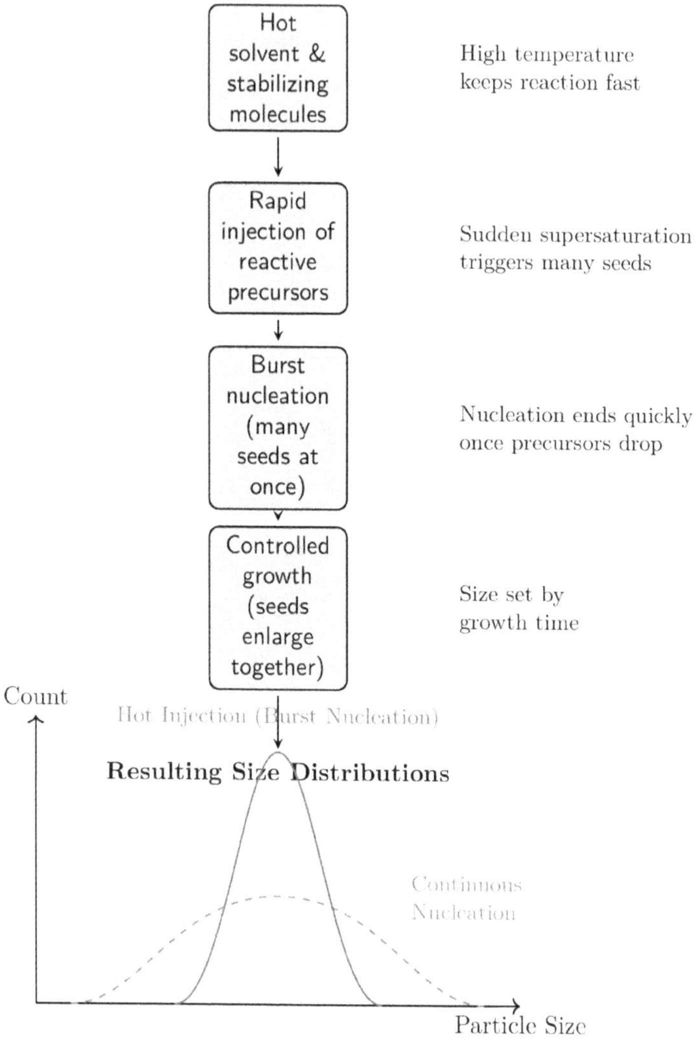

Narrow size spread ⇒ sharper emission spectra

In hindsight, this may sound like the obvious way to proceed: synchronize birth, then manage growth. But "obvious" is often what becomes obvious *after* someone demonstrates that it can work in a real flask, with real reagents, against all the ways chemistry likes to misbehave. Colloidal nanocrystals are made in a regime where competing processes are always waiting to spoil the party: particles can fuse, dissolve and re-precipitate, exchange material with one another, or grow in fits and starts when precursors arrive unevenly. A recipe that makes these tendencies line up into a repeatable outcome deserves to be called a breakthrough.

The Nobel committee's later language about 1993 being the moment production was "revolutionised" toward "almost perfect" particles may sound ceremonial, but it points to a technical truth. The most valuable product of the synthesis was not any particular color. It was reliability: a way to make samples whose spectra were not vague averages over a disorderly population.

Why was monodispersity so transformative? Because it turned optical measurements into a kind of metrology. When particles are narrowly distributed, the absorption features and photoluminescence peaks become sharper. Sharp peaks can be tracked from batch to batch, and small differences become meaningful rather than lost in the noise of polydispersity. A spectrum stops being just a fingerprint and becomes a ruler. It can tell you, quickly, whether a synthesis produced the size you aimed for, whether the growth time was too long, whether something unexpected happened at the surface.

There is also a psychological threshold here. Engineers and product designers do not like miracles; they like tolerances. A display manufacturer cannot build a supply chain around a material that "usually" emits green. They need a material that emits green with a defined peak wavelength and a defined linewidth, and that does it the same way next month as it does today. A narrow distribution is, at bottom, a way of turning a lab curiosity into a component that can be specified, tested, and purchased.

The 1993 approach also improved crystallinity, which is a word that can sound abstract until you connect it to what photons and excitations "see." In an imperfect nanocrystal, internal defects can provide pathways for energy to disappear as heat rather than light. Better crystallinity generally means fewer such pathways, and therefore brighter, more stable luminescence. Even when the earliest technological ambitions were not yet in view, a brighter sample had immediate scientific consequences: stronger signals allow faster experiments, subtler comparisons, and clearer trends.

Anecdotes from laboratories at the time often describe the almost visceral difference between "before" and "after." In a typical earlier synthesis, a researcher might run a reaction, measure a broad absorption feature, and spend days debating whether the shoulder on the curve was meaningful or merely the sign of a mixture. After the hot-injection strategy began to spread, it became possible—sometimes after a few hard-won iterations—to generate a series of samples where changing the growth time produced a predictable march of spectral peaks. That does not sound like drama, but it is the drama of control replacing guesswork. A graduate student's week becomes less about hoping the reaction behaves and more about using it as an instrument.

The method's setting also mattered. Doing this chemistry in a hot coordinating solvent provided a medium that could dissolve precursors, permit rapid reactions, and supply molecules that bind to the nanocrystal surface. Those surface-bound molecules perform a quiet but essential role: they help keep particles from sticking together and can moderate growth by controlling which atomic sites are easily accessible. It is tempting to treat such molecules as mere "stabilizers," but the deeper point is that the surface is not an afterthought. At nanometer sizes, the surface is a governing part of the system, and chemical control of that boundary is one of the main ways the material becomes reproducible.

A striking continuity links this to the earlier glass work. The glass matrix constrained particles physically, preventing them from drifting together and coalescing; in solution, the coordinating solvent and surface-bound molecules provided a chemical analogue of that constraint. The mechanisms differ—rigidity versus molecular binding—but the goal is closely related: keep particles distinct and give growth a controlled tempo. Retrospective profiles have noted this conceptual analogy, and it helps explain why the overall story reads as a chain rather than a set of disconnected tricks.

Hot-injection also clarified a subtle point about what it means to "tune" a nanocrystal. Tuning is often described as simply choosing a size, but size is not a single number. Real nanocrystals have shapes, facets, and surfaces with different atomic terminations. Two samples with the same average diameter can behave differently if one is more faceted, more defective, or more chemically rough at the boundary. The 1993 synthesis did not solve all of these problems at once, but it moved the field closer to treating nanocrystals as *objects* with controllable attributes rather than as vague "small particles."

From that perspective, the most consequential change was the emergence of a new experimental rhythm. Once nearly monodisperse samples could be produced, researchers could stop fighting their materials and start interrogating them. Questions that had been muffled by polydispersity became audible: How does the luminescence linewidth depend on size? How stable is the emission under continuous illumination? How does modifying the surface change brightness or photostability? When you can make batches that are similar, you can change one variable at a time—and "one variable at a time" is the heartbeat of mature experimental science.

A second consequence was communal. A recipe that can be followed and adapted is a vehicle for standardization. If a result depends on a uniquely skilled practitioner, it remains fragile. If it can be replicated by groups with different backgrounds and different equipment, it becomes part of the shared toolkit. The Murray–Norris–Bawendi synthesis gave the community a common starting point: a method that could be tweaked, improved, diversified, and, crucially, taught.

That teachability is not a soft virtue. It is how a technology becomes scalable without losing its scientific core. When a method becomes reliable enough to spread, it creates a feedback loop between basic research and practical ambition. Better syntheses enable better measurements; better measurements reveal which imperfections still matter; that knowledge drives improved syntheses. Over time, the material becomes not only more uniform in size but more uniform in behavior—brighter, more stable, more predictable.

It is worth pausing on a practical image that captures the change. Imagine a tray of small vials, each holding a colloidal dispersion. Under ordinary light, they may look only faintly tinted. Under ultraviolet light, they glow in distinct colors. Early on, such a tray might look charming but unreliable: colors overlap, brightness varies, and the same reaction repeated a week later might produce a noticeably different palette. With improved control, the tray becomes a calibrated set. Each vial corresponds to a chosen size, and the colors are separated cleanly enough that you can imagine building with them—printing them into films, embedding them in plastics, integrating them into devices where consistency is not negotiable.

That is what a "platform" looks like in chemistry. It is not a single compound, but a way of making a family of related materials with predictable properties. Once the platform exists, applications stop being speculative. A display engineer can ask for a particular emission wavelength and a particular linewidth. A biologist can ask for a label that emits at a wavelength that avoids background fluorescence.

A physicist can ask for samples narrow enough in size that collective measurements reveal intrinsic behavior rather than a smeared average.

The 1993 recipe did not end the story of synthesis—it began the era in which synthesis became the story's engine. It offered a practical answer to the question that had been hovering since the first seductive spectra: can this be made on purpose, again and again, with properties that are not merely observed but chosen?

Once the answer became "yes," the glowing nanocrystal stopped being a lucky artifact at the bottom of a beaker. It became something closer to a manufactured object, small enough to behave in quantized steps and consistent enough to be treated as a component—an ingredient that could be ordered, tested, and trusted.

4.4 Why Stockholm Honored a Chain, Not a Moment

Nobel Prizes are often imagined as medals pinned to a single flash of insight: a scientist sees what nobody else has seen, the world rearranges itself, and history obligingly supplies a date. Real technologies rarely cooperate with that storytelling. They arrive the way cities do—through foundations that vanish underground, through scaffolding that looks temporary until it becomes permanent, through thousands of ordinary decisions that make an idea sturdy enough to carry weight.

That tension was built into the 2023 Nobel Prize in Chemistry. The committee did not reward a lone "eureka" but a long relay: Alexei Ekimov's early observation of size-dependent effects in glass in the early 1980s; Louis Brus's demonstration and model-driven understanding of size-dependent effects in colloidal systems in 1983–

1984; and Moungi Bawendi's role in the 1993 hot-injection synthesis that made high-quality, nearly monodisperse nanocrystals a reproducible material. The official rationale framed the arc as discovery, understanding, and chemical control—three links that had to lock together before a laboratory curiosity could become something the world would quietly come to depend on.

That framing is more than a diplomatic compromise. It reveals how modern science advances when a phenomenon is both subtle and useful. Subtle phenomena can be discovered by accident, but they cannot be industrialized by accident. And usefulness is not a property of an effect alone; it is a property of an effect plus a method for making it reliably.

Consider what "embedded in daily life" actually means in this case. A mature technology does not announce itself with a name tag. People do not generally say, "I spent the afternoon watching nanocrystals." They say, "The colors on that screen look good," or "That medical test caught something early," or "That label is bright enough to see." Technologies like these succeed by becoming ordinary.

Displays are the easiest place to feel this ordinariness. Most people have seen a television advertised with richer reds and cleaner greens. Behind that marketing language sits an engineering problem: how to produce narrow, well-placed colors. Many display architectures begin with a broad-spectrum light source and then filter it into red, green, and blue components. Filters are wasteful: they throw away light you paid to generate. A material that can convert part of a broad spectrum into narrow bands—especially in the red and green where human vision is sensitive—acts like a more economical kind of filter, closer to a translator than a sieve. The widespread use of nanocrystal-based color conversion in consumer displays is not a triumph of esoteric physics; it is a triumph of predictable manufacturing meeting a stubborn engineering constraint.

Lighting and sensing follow the same pattern. In lighting, what matters is efficiency, stability, and controllable spectra. In sensing, what matters is signal-to-noise: a tag or probe that emits in a tight band can be separated cleanly from background light and from other tags. In biomedicine, those tight bands enable multiplexing—several labels used at once without turning the measurement into a blur. The point is not that every application depends on exactly the same material or device design; the point is that all these applications rely on the same underlying promise: a tunable optical response that can be produced repeatedly, with controlled quality.

That promise required three different kinds of expertise, and that is another reason Stockholm honored a chain.

The first expertise was observational, rooted in solid-state optics and materials. Ekimov's glass work in the early 1980s offered a stable environment where nanocrystals could form and be measured without immediately aggregating or degrading. Glass acted as both host and constraint: it let nanocrystals exist long enough to show their strange optical behavior. The key contribution here was not yet a manufacturing recipe for mass production. It was an empirical shock: a material's fundamental optical threshold could shift because the crystals were tiny.

The second expertise was conceptual and predictive, rooted in chemical physics. Brus's early work in 1983–1984 did something that often marks the difference between an observation and an accepted principle: it connected the size dependence to a framework that could be used to anticipate trends and link them to other measurable quantities. In Brus's hands, the phenomenon was not limited to a particular glass recipe or a particular lab's furnace schedule. It became a general statement about small semiconductor crystallites, including how their energies could shift in ways that were chemically meaningful, such as through changes in redox potentials. That step is easy to underappreciate if one thinks of models as ornamentation. A good model is a discipline: it tells a community what counts as plausible, what counts as surprising, and what experiments are worth doing next.

The third expertise was synthetic mastery, rooted in practical chemistry. The 1993 hot-injection work associated with Bawendi and colleagues changed the daily reality of the field. Nearly monodisperse samples made optical data sharper and made "size tuning" something that could be specified rather than hoped for. That shift is exactly what turns a discovery into a platform. If the discovery is the first time someone sees the behavior, the platform is the moment the behavior becomes available to others as a material they can hold, measure, and integrate into devices.

When the Nobel committee emphasizes an "arc," it is pointing toward a broader lesson about how physics and chemistry divide labor in technology. Physics is often where the phenomenon is first made legible: it supplies the concepts that tell you what to look for and why it might happen. Chemistry is often where the phenomenon becomes consequential: it supplies the methods to make the material reproducibly, to tune it, to stabilize it, and to scale it. This is not a hierarchy. It is an interdependence. The nanocrystal story is a case where the core effect sits at the border of the disciplines, and the world-changing part lies in the bridge between them.

A simple analogy helps. Imagine a musical instrument that produces a beautiful, unfamiliar note, but only when held at exactly the right angle, in exactly the right humidity, by one unusually patient performer. That is a discovery: the note exists. A technology begins when the instrument can be built in a factory so that many musicians can play the note reliably on a stage. The Nobel committee's chain is, in effect, the chain from hearing the note to building the instrument.

There is a human side to this chain that rarely shows up in the tidy retrospective. Research cultures have their own dialects. A group working on glass-hosted nanocrystals speaks in terms of heat treatments, phase separation, and solid matrices. A group working on solution-phase nanocrystals speaks in terms of precursors, solvents, ligands, purification, and colloidal stability. A group building devices speaks in terms of films, encapsulation, photostability, and manufacturing tolerances. The same physical object can appear in all three worlds, but it is described differently, valued differently, and tested differently. A chain of discoveries is also a chain of translations.

That translation work is part of why timeline discussions can become contentious. Scientific communities are understandably sensitive to priority: who saw what first, who interpreted it correctly, who made it usable. High-reputation reporting around the 2023 prize added context and also cautioned that timelines can be misreported and later corrected. The Nobel materials, anchored in the primary literature and the committee's deliberations, served as the stable narrative spine: early observation in glass, solution-phase establishment and understanding, and a synthesis breakthrough enabling high-quality controllable materials. That is not a claim that no other contributors mattered; it is a claim about the essential links that made the effect both real and useful at scale.

That distinction—between "who contributed" and "what links were necessary"—is the philosophical heart of the prize. Nobel committees are not awarding a census. They are awarding a causal story about how an intellectual tool entered the world.

Manufacturing reliability, in this view, is not merely an engineering add-on. It is intellectually significant. A reproducible synthesis does not just produce more material; it makes the phenomenon testable in deeper ways. With uniform particles, differences between theory and measurement become interpretable rather than dismissible. With consistent batches, the effect can be explored across temperatures, environments, and device structures. Reliability is a kind of clarity. It is what allows a community to stop arguing about whether an effect exists and start asking how far it can be pushed.

This is also where the moral texture of the technology emerges. Once a material becomes manufacturable, society inherits its benefits and its risks. Semiconducting nanocrystals have raised environmental and toxicity concerns in some compositions, prompting efforts toward safer alternatives, improved encapsulation, and responsible life-cycle thinking. Those concerns are not separate from the Nobel story; they are part of what it means for a nanoscale phenomenon to leave the laboratory and enter commerce. A chain that ends in manufacturing ends, unavoidably, in questions about stewardship.

The 2023 prize also carried a quieter message about modern "chemistry" itself. Chemistry is sometimes caricatured as the science of mixing and measuring, a pragmatic craft compared with the supposedly purer abstractions of physics. Yet the nanocrystal saga highlights a different truth: chemistry is the discipline that decides what kinds of matter can be made with intention. It is a science of *control*. In that light, a synthesis method that yields nearly uniform nanoparticles is not just a clever procedure; it is a way of shaping the options available to everyone else—physicists who want clean systems, engineers who want reliable parts, and biologists who want bright labels that behave consistently in messy environments.

In the end, Stockholm honored a chain because a chain is what carried the idea into the world. A single observation can be astonishing, but astonishment is not a supply chain. A model can be elegant, but elegance does not fill a vial on demand. A synthesis can be ingenious, but ingenuity becomes transformative only when it can be repeated, taught, and trusted.

When those links lock together, the nanoscale stops being merely small. It becomes *choosable*. And once a property becomes choosable—once color becomes something you can dial in by design rather than accept as fate—matter starts behaving less like a fixed inventory and more like an editable medium.

Chapter 5

Cooking Light in a Beaker: The Chemistry of Making Perfect Dots

A quantum dot is less like a tiny bead than like a performance staged at high temperature, where a few seconds can decide whether a flask fills with identical emitters or a muddle of mismatched grains. In these pages, the reader watches color being cooked into matter itself and discovers why the most important part of a glowing crystal may be the molecules clinging to its surface rather than the crystal alone.

5.1 Hot-Injection: The Moment a Nanocrystal Is Born

A good quantum dot synthesis has the emotional profile of a suspense scene: long preparation, then a sudden event that cannot be taken back. The flask sits at temperature, apparently calm. The solution may be clear, perhaps faintly colored by the solvent or ligands. A syringe waits with precursors that are stable enough at room temperature to be handled, but reactive enough to become a semiconductor once they meet heat and the right chemical partners. When the injection happens, a liquid that looked inert becomes a factory for crystals so small that their color will later be tuned by fractions of a nanometer.

The crucial idea is not simply that injection starts the reaction. It is that injection can concentrate the reaction into a brief, intense moment, and that this moment is the chemist's best chance to keep the final particles uniform. Uniformity is the difference between a clean spectral line and a smeared rainbow; between a display film that gives a precise shade of green and one that wastes light in unwanted wavelengths; between a bioimaging probe that pinpoints a tumor and one that blurs into background.

To see why, it helps to think the way colloid chemists learned to think about monodispersity long before anyone said "quantum dot." LaMer and Dinegar (1950) described a now-classic picture of how a cloud of identical particles can emerge from solution. The ingredients that can become the solid (often called "monomers," even though they are not monomers in the polymer sense) build up in concentration. Nothing happens at first; the solution can hold them in a metastable state. Then, at a threshold of *supersaturation*—meaning the solution contains more of the would-be solid than it can comfortably tolerate—nuclei suddenly appear. After that, the monomer concentration drops as the nuclei grow, and the system settles into a gentler regime where existing particles get larger rather than new ones being born.

Real nanocrystal syntheses are messier than the textbook cartoon. Precursors convert at finite rates; side reactions occur; clusters can form and reorganize; ligands bind and unbind. Yet the LaMer-style logic remains powerful as a target: pack nucleation into a narrow time window, then feed growth onto an existing population. Hot-injection is a way to push the chemistry in that direction by using heat and mixing as a switch.

The switch has two parts. The first is chemical: precursors that are sluggish at room temperature can react quickly at a higher temperature, forming the building blocks of the crystal at a rate that briefly overwhelms the solution. The second is physical: injecting a small volume of concentrated precursor into a larger volume of hot coordinating solvent can create, for a short time, a high local concentration of reactive species—high enough that nucleation is favored everywhere in the flask, not only near the injection point.

That "everywhere" is aspirational. In practice, the earliest milliseconds can be the most dangerous. If mixing is imperfect, one region of the flask can experience a higher supersaturation than another. That region nucleates first, consuming monomer and lowering the driving force for nucleation elsewhere. The result is not one birth event but several staggered ones, and staggered births are the seed of a broad size distribution: older particles start growing earlier and have a head start that is hard to erase.

This is why chemists who make nanocrystals talk like experimentalists and like engineers at the same time. A synthetic protocol is not just a set of reagent amounts; it is an orchestration of heat flow, fluid flow, and reaction kinetics. The same nominal recipe can give different results in a 25 mL flask and a 2 L reactor because the mixing time, thermal mass, and heat transfer coefficients are different. Even in a small flask, the angle of the needle, the speed of injection, and the vigor of stirring can be the difference between "nearly monodisperse" and "good enough for a lab demo."

A historical example made this visceral for a generation of researchers. In 1993, Christopher Murray, David Norris, and Moungi Bawendi reported nearly monodisperse CdE nanocrystals (E = S, Se, Te) made by injecting organometallic precursors into a hot coordinating solvent. That paper did not merely add a new material to the world; it made a promise: if nucleation can be forced into a burst, growth can be treated like a controlled enlargement of a well-defined population. It is difficult, now, to recapture how transformative that felt. Before that, “small semiconductor particles” often meant poorly defined clusters with broad optical features. The hot-injection approach offered something closer to a synthetic instrument: turn a knob—temperature or time—and get a predictable change in size and therefore optical properties.

The promise is real, but it rests on a delicate physical bargain. Nucleation must be brief not because brevity is aesthetically pleasing, but because nucleation and growth compete for the same feedstock. The feedstock is the pool of monomeric building blocks (or small reactive fragments) that can either assemble into brand-new nuclei or attach to existing nanocrystal surfaces. If new nuclei continue to appear while older particles are already growing, the system is trying to do two incompatible tasks at once: increasing the *number* of particles and increasing their *size*. The monomer pool is finite, so each task starves the other in uneven ways, and the resulting population becomes a mixture of ages, sizes, and often shapes.

A quick way to feel the logic is to imagine two particles that differ only by being born at different times. The older one experiences a higher monomer concentration for longer, so it grows larger. If the younger one is born late, after the monomer concentration has already been lowered by growth, it may never catch up. The final sample can easily become a blend: a small fraction of larger, early-born particles that dominate absorption, and a larger fraction of smaller, late-born particles that broaden the emission. The eye sees this as dullness: color that refuses to be pure.

Temperature turns out to be the main lever for keeping nucleation brief, but it does so in more than one way. Higher temperature tends to speed precursor conversion, increasing the rate at which monomer appears. It can also lower the effective barrier for nucleation, meaning the solution requires less supersaturation to begin forming nuclei. These effects pull in opposite directions with respect to timing: faster monomer generation pushes the system quickly above the nucleation threshold, but an easier nucleation threshold can allow nucleation to start earlier and last longer. Meanwhile, temperature also affects growth: surface attachment, diffusion through the ligand shell, and internal rearrangements can all be accelerated by heat. That is why a reaction temperature that gives beautiful dots at one precursor concentration can produce a mess at another.

The coordinating solvent and ligands are not passive spectators in this drama. A "coordinating solvent" is one whose molecules can bind to the surface atoms of a growing nanocrystal. This binding can stabilize small clusters that would otherwise dissolve, and it can slow growth by occupying surface sites. It can also change precursor reactivity by binding to metal centers or by altering the local environment where precursor conversion happens. A ligand shell, even at the earliest stages, changes the effective surface energy of the nuclei. Lower surface energy tends to make nucleation easier, which can be a blessing or a curse: it can sharpen the burst by allowing many nuclei to appear quickly, or it can prolong nucleation by allowing even small fluctuations in monomer concentration to generate new nuclei for longer than intended.

The atmosphere over the flask matters for similar reasons. Many precursor systems are sensitive to oxygen or water, which can consume reactive intermediates or create oxide defects. Even when contamination does not destroy the synthesis, it can subtly change kinetics. A trace of water can protonate a ligand, shifting binding equilibria and changing how strongly the surface is passivated. A trace of oxygen can oxidize a precursor, changing its conversion rate and therefore the timing of supersaturation. The result can be a sample that looks "almost right" by eye but has a broader spectrum, more trap states, or poorer stability—failures that may only show up later when the dots are processed into a film or placed in a biological medium.

There is a second subtlety in the nucleation—growth bargain: the supply of monomer does not merely rise and fall; it can be throttled. Some syntheses are "precursor-limited," where monomer appears gradually because precursor conversion is slow. Others are "growth-limited," where monomer appears quickly but growth is slowed by surface chemistry. Hot-injection was originally celebrated because it could create a high, sudden monomer concentration that triggers nucleation, followed by depletion that naturally ends nucleation. But the same technique can fail if the precursor continues converting at a high rate after injection. In that case, monomer does not deplete; it stays high; nucleation can continue; growth can become uncontrolled. What looked like a single birth event becomes a prolonged maternity ward.

In the lab, these distinctions show up as anxieties about timing and heat. A few seconds can matter because seconds can be the difference between a high supersaturation pulse and a long plateau. Stirring speed can matter because it determines whether the injection creates a uniform concentration spike or a localized plume. The size of the needle bore can matter because it controls how quickly the precursor volume enters, which in turn controls the sharpness of the supersaturation pulse. The same is true of the solvent temperature at the moment of injection: if the injection cools the solvent noticeably, precursor conversion slows, supersaturation rises more slowly, and nucleation can smear out over time.

Reproducibility, then, is not a moral virtue; it is a measurement problem. The chemist has to know what the temperature *actually* is in the region where nucleation occurs, not just what the hot plate reads. The chemist has to know how quickly mixing happens relative to the onset of nucleation. In a small flask, it is tempting to assume mixing is instantaneous, but "instantaneous" is an engineering statement, not a feeling. Milliseconds to seconds are long enough for nucleation to begin in a plume and set the population trajectory.

A helpful way to keep the logic straight is to treat the reaction as having a few chemically meaningful phases, even if the boundaries blur in real time. First comes accumulation: monomer-forming reactions turn on, and the concentration climbs. Then comes the nucleation window: a brief period where nuclei form rapidly because supersaturation is high. Then comes depletion and growth: monomer concentration drops, nucleation becomes unfavorable, and existing particles grow by consuming the remaining monomer. If the system is well designed, that growth can be controlled—either by letting it proceed at a steady rate or by deliberately limiting monomer supply to avoid runaway enlargement or shape changes.

If the goal is a narrow distribution, the most important creative act is not the slow growth that follows but the timing of birth. One can treat it like casting a theater performance. If the casting call stays open for hours, the ensemble will contain actors of different ages and experiences, and they will not move in synchrony. If the casting call is brief and decisive, everyone starts rehearsals together. Growth is rehearsal; nucleation is casting.

Chemists therefore put obsessive attention into the moment of injection. The precursor solution is often prepared to be as homogeneous as possible. Syringes and needles are chosen to deliver a reproducible flow. The receiving solvent is heated sufficiently above the intended growth temperature so that the temperature dip from injection does not shove the system into a different kinetic regime. In many labs, the "real" growth temperature is reached only after injection, once the system re-equilibrates and the monomer concentration has already begun to fall. This can look like fussy choreography, but it reflects a hard truth: the physical process of getting molecules from a syringe into a flask is inseparable from the chemical process of making nuclei.

This is also why scale-up is notoriously difficult. A synthesis that works beautifully in a small flask often relies on rapid mixing and fast heat recovery. In a larger vessel, injection may create jets and gradients; stirring may not homogenize the volume quickly enough; the temperature may not be uniform. The system can slip from "burst nucleation" to "continuous nucleation" without any change in nominal concentrations. The final product may still glow—quantum dots are forgiving in that sense—but the glow becomes broader, and broad glow is the enemy of both color purity and quantitative sensing.

The phrase "tiny slips" in timing or temperature understates the consequence, because the sensitivity is often exponential. Nucleation rates can depend very steeply on supersaturation: a modest change in concentration can trigger an enormous change in how many nuclei form per unit time. Precursor conversion rates can depend steeply on temperature: a few degrees can make the difference between a sharp pulse of monomer and a slow leak. A synthesis can appear stable for several runs and then, with a slightly different reagent batch or a slightly different ambient humidity, drift into a regime where the nucleation window widens. The result is not a gentle degradation but a sudden broadening of the distribution that no amount of extra growth time can repair.

There is a final twist that makes hot-injection feel almost philosophical. The burst is, in part, a deliberate act of scarcity. The system is engineered so that nucleation quickly consumes the "freedom" to nucleate by depleting monomer below the threshold. After the burst, the chemistry is biased toward growth on existing surfaces, not because growth is intrinsically preferred, but because the system has been pushed into a state where only growth is kinetically accessible. In other words, uniformity is purchased by forcing the system out of its most creative phase as quickly as possible.

When it works, the reward is striking: a flask that, under ultraviolet light, shines with a single, saturated hue, hinting that most particles are the same size and therefore share nearly the same electronic transition energy. That hue is not just a property of a material but the fingerprint of a moment—the brief interval when supersaturation crested, nuclei appeared almost together, and the monomer pool was then pulled back down before the solution had time to keep inventing new crystals.

5.2 Size Control Is Color Control

After the burst of birth, the flask enters a quieter phase that feels almost domestic: hold the temperature steady, keep stirring, wait. Yet this waiting is where light is tuned. A nanocrystal that gains a few layers of atoms is not simply "more of the same material." As was already clear from the size–bandgap rule, changing size changes the effective gap between allowed electronic energies. In practice that means growth is not just the path to "more product"; it is the dial that selects a hue.

Many chemists first meet this fact in the most concrete way possible: by stealing tiny samples from a running reaction. An experienced hand will withdraw an aliquot with a pipette, squirt it into a cold solvent to halt growth, and shine a UV lamp on the vial. Early samples glow toward the blue-green end, later ones drift toward yellow, orange, then red. The colors are not paint mixed in a palette; they are the optical consequence of nanocrystals crossing particular sizes. It is hard not to find that a little uncanny: time in the flask becomes a visible gradient.

CdSe has been the classic teaching material for this reason. In the early days of colloidal nanocrystal synthesis, one could almost do spectroscopy by eye. The size series demonstrated by Murray, Norris, and Bawendi (1993) made the point unavoidable: hold everything else reasonably similar, and just by letting growth continue at temperature, the absorption edge and emission color slide systematically. A student can watch the reaction "redden" and, without knowing any equations, learn a discipline that later matters for televisions and medical imaging: if you want the same color tomorrow that you made today, you must control the conditions that control size.

The everyday knobs are straightforward to list, but not to master: temperature, time, precursor ratios, and how quickly fresh building blocks are supplied. Each of these knobs touches the same underlying variable, the *monomer concentration* in solution, and that concentration is what the nanocrystal surfaces feel as a growth pressure.

Temperature is the bluntest instrument. Higher temperature usually accelerates the conversion of precursors into reactive monomer and speeds the microscopic steps of growth on the crystal surface. That can push nanocrystals through size ranges quickly, which is useful when a particular size must be reached before ripening or shape changes become a problem. But high temperature can also produce side reactions, degrade ligands, or keep monomer concentration high enough that the system never fully stops nucleating. The optical penalty arrives as broad spectra: too many sizes present at once, so too many emission energies.

Time is the instrument every chemist is tempted to treat as simple. "Longer time means larger particles" often works as a first approximation, but it hides two complications. The first is that growth rate usually slows down: early on, monomer may be abundant, surfaces highly reactive, and the driving force strong; later, monomer is depleted and the surfaces are increasingly protected by ligands. The second is that the population can change its *distribution* over time in ways that are not monotonic. A reaction can begin broad, become narrower, and then broaden again if conditions drift into a ripening regime. Watching only the peak color can be misleading; watching the width of the peak is what tells you whether control is being gained or lost.

Precursor ratios matter because they determine what "monomer" means chemically. In a II–VI semiconductor such as CdSe, the crystal requires cadmium- and selenium-containing fragments in the right balance to build the lattice. If one precursor is effectively limiting, growth can become bottlenecked by that element, altering both kinetics and the likelihood of defects. If one precursor is in large excess, it can saturate the surface with a particular type of ligand-like fragment, changing which facets grow fastest and thereby nudging shape as well as size. Ratios also change the speed and timing of monomer generation, so they feed back into whether nucleation truly ended when it was supposed to.

The most instructive knob, conceptually, is *feeding*: whether monomer is delivered as a one-time pulse (as in classic hot-injection) or as a controlled trickle (as in various "continuous injection" and related strategies used in many labs). Feeding turns growth into something closer to baking than to fireworks. If monomer is supplied too quickly, concentration rises, and the system may cross back into a nucleation-favorable regime, spawning new small particles that will later pollute the spectrum. If monomer is supplied slowly, concentration can be held in a sweet spot where existing particles grow but new nuclei are unlikely. This is one of the reasons colloidal nanocrystal synthesis resists being reduced to a recipe: the *rate* at which molecules are made available can matter as much as the *amount*.

All of these knobs ultimately confront a central worry: dispersity, the spread of sizes in the sample. Dispersity is not just a nuisance for perfectionists. Because optical spectra respond directly to size, dispersity becomes visible as peak broadening. A narrow distribution produces sharp excitonic features; a broad distribution washes those features out into a fat hump. A spectrometer is therefore a kind of honesty machine for synthetic control: the line shape reports what the size distribution is doing, whether or not the solution looks pretty.

Why does the distribution change at all once nucleation is over? The naive expectation is that if every particle grows by adding the same amount of material, the distribution simply shifts to larger size without changing shape. That is not what happens, because growth rates can depend on size. If smaller particles grow faster than larger ones, the distribution *focuses*: it narrows as time goes on, as if the stragglers are catching up. If larger particles grow faster, the distribution *defocuses*: the rich get richer, the distribution spreads, and optical peaks broaden.

Xiaogang Peng and coworkers (1998) gave this behavior a clear kinetic vocabulary, framing focusing and defocusing as outcomes of how monomer concentration interacts with size-dependent growth rates. The key intuition can be stated without mathematics: surface curvature and surface chemistry make small particles behave differently from large ones. Small particles have higher surface energy and often a greater thermodynamic "hunger" to grow, but whether that hunger turns into faster growth depends on whether monomer can reach and attach to their surfaces under the ligand shell. At high monomer concentration, attachment may be so easy that large particles—having more surface area—capture more monomer and accelerate ahead. At moderate concentration, small particles may have a relative kinetic advantage and narrow the distribution. The same reaction can pass through both regimes as monomer is consumed, which is why a sample can tighten up and then fall apart.

This dynamic quality makes the idea of "the right growth time" slippery. Suppose a synthesis begins with a distribution that is not perfectly narrow—almost always the case. There may be a time window where focusing is strongest and the distribution is at its narrowest. Past that window, if monomer concentration drops too low, the system can slide into ripening.

Ostwald ripening is the term for a tendency that sounds cruel but is simply thermodynamics wearing a lab coat: smaller particles dissolve and larger particles grow. The driver is surface energy. Atoms on a highly curved, small particle are less stable than atoms on a flatter surface, so they are more likely to detach into solution. Once detached, they are more likely to end up on a larger particle, where they are more stable. Ripening can be slow or fast depending on temperature, solvent, ligand binding strength, and whether the system contains any residual reactive precursor that can replenish monomer. If ripening dominates, the average size can increase while the distribution broadens, and the color may continue to redden while the spectral peak becomes disappointingly wide.

Ripening has another, subtler effect: it can change shape. Growth is not obliged to preserve a spherical form. Different crystal facets have different surface energies and different ligand coverages, so they can grow at different rates. Under some conditions, a particle that began roughly spherical can evolve into a rod, a platelet, or another anisotropic form. Peng and Peng (2000) discussed mechanisms by which CdSe nanocrystals change shape as growth conditions change, emphasizing that "growth" can include internal reorganization and facet competition, not only the outward addition of atoms. For the optical properties, this introduces a complication: size is no longer the only variable. Shape affects how electrons and holes are confined, and it affects how the nanocrystal interacts with light through polarization and selection rules. A synthesis that seems to control size may still drift in optical behavior if it also drifts in shape.

These are not merely academic subtleties. A display manufacturer does not buy “nanocrystals”; it buys a particular emission peak at a particular wavelength with a particular linewidth, stable under heat and blue-light pumping. A biomedical researcher does not buy “a red-emitting dot”; they buy a probe that remains bright in salty water, whose emission is distinct from tissue autofluorescence, and whose spectrum can be separated cleanly from other labels. In both cases, linewidth is precious because it allows separation: separate colors on a screen without crosstalk; separate multiple biological markers without spectral overlap.

The chemist’s controls therefore become optical promises. Take temperature again, but now seen through linewidth. If the temperature is too low, precursor conversion may be sluggish, growth may be uneven, and the sample may contain a mixture of sizes reflecting uneven access to monomer. If the temperature is too high, ligands may desorb too readily or decompose, exposing reactive surfaces that can fuse or ripen; nucleation may restart; the distribution spreads. There is often a narrow band of temperatures where growth is fast enough to be practical but gentle enough to preserve monodispersity.

Precursor concentration and feeding strategies also translate into optical outcomes. A one-shot growth that starts with a high monomer concentration might push the sample rapidly toward the desired size, but it can also favor defocusing if larger particles capture monomer more efficiently. A controlled feed can keep monomer concentration in a focusing regime longer, allowing a sample to “self-correct” toward uniformity. The desired result is not only the right peak position but a tight peak width, because that tightness is the signature of a disciplined population.

That phrase—*population*—is worth lingering on. Nanocrystal synthesis is not like synthesizing a small molecule, where the goal is to make identical copies of a structure. It is closer to agriculture. One does not harvest a single plant; one harvests a field whose distribution of sizes, health, and maturity depends on weather, soil, and timing. The chemistry sets the climate; the ligands are the soil; the monomer concentration is the rainfall; temperature is the season. A skilled chemist learns to think in distributions and to treat spectra as the census report.

A concrete laboratory moment captures how unforgiving this can be. Imagine two flasks run nominally the same way, except that in one, the injection took five seconds longer because the precursor solution was slightly more viscous, or the needle was narrower, or the operator hesitated. That small delay can broaden the nucleation window by allowing monomer to build gradually rather than sharply, which leaves the growth stage starting with a wider age distribution. Once the growth stage begins, focusing might narrow the spread somewhat, but it cannot fully erase an age difference that translates into size advantage. The later optical result may be a peak that is slightly red-shifted (because some particles grew larger early) and noticeably broader. If those dots are used as a color-conversion layer in a display backlight, that broadening can wash out the color gamut; if they are used in a multiplexed bioassay, it can increase channel overlap and reduce sensitivity.

The temptation is to treat these outcomes as minor imperfections. After all, the solution still glows. But the modern uses of nanocrystals reward precision. A narrow peak is not a luxury; it is a way of using photons efficiently. Light emitted in unwanted wavelengths is light that must be filtered or is simply wasted. In a device, wasted photons become heat. In tissue, wasted photons become background.

The deep connection between size control and optical control is one reason the field advanced so quickly once reliable syntheses existed. A new synthetic trick could be evaluated immediately by spectroscopy, without waiting for a device to be built. If the first excitonic absorption peak sharpened, if the emission linewidth narrowed, if the peak position could be dialed smoothly by growth time, the synthesis was doing something right. Spectra became the feedback loop that let chemists turn reaction conditions into a reproducible palette.

At the end of a successful growth, there is a particular satisfaction in seeing a sample whose spectrum looks almost like a single note. The color is not merely "red" or "green," but a specific shade, repeatable, with edges that feel crisp. The nanocrystals have become, in a practical sense, quantized pigments: their size distribution has been disciplined tightly enough that a macroscopic object—a vial of liquid—behaves as if it contains one carefully chosen electronic transition, multiplied by trillions of identical copies.

5.3 Ligands: The Molecular Handshake at the Surface

A freshly grown nanocrystal has an awkward secret. The interior may resemble a tidy chunk of semiconductor, but the outside is a cliff edge: atoms at the surface have fewer neighbors than atoms in the middle, and those missing neighbors translate into unsatisfied bonding possibilities. In everyday solids that hardly matters, because the fraction of atoms on the surface is tiny. At the nanoscale it becomes a defining fact. The particle is so small that a large share of its atoms live at the boundary, and the boundary is chemically restless.

That restlessness has two consequences that chemists learned quickly. First, bare nanocrystals stick to each other. If you take many tiny particles with reactive surfaces and put them in a solvent, they tend to collide and fuse or clump, because sticking lowers the total surface area and therefore lowers surface energy. The second consequence is optical: those unsatisfied bonding possibilities create electronic states that sit in the wrong place energetically. They can trap charges or excitations and route the energy into heat rather than light. A nanocrystal that should glow can become dim, fickle, or maddeningly intermittent.

Ligands are the social solution to this antisocial surface. The word comes from coordination chemistry: a ligand is a molecule (or ion) that binds to a metal center. On a nanocrystal, ligands bind to surface atoms in the same general spirit. They do not merely float nearby; they attach. Their head groups (the binding ends) coordinate to cadmium, zinc, lead, indium, or whatever atoms are exposed. Their tails (often hydrocarbon chains) reach out into the solvent. In a good synthesis, ligands arrive early and in abundance, so the nanocrystal is never truly naked. It is born wearing an organic coat.

That coat does several jobs at once, which is why a discussion of ligands can feel like trying to pin down a talented multi-instrumentalist. Ligands keep particles apart in solution. They help decide which solvents the particles like. They influence how fast the particles grow and what shapes they adopt. They can prevent the worst surface electronic states from forming, raising brightness. They also create headaches: the same coat that makes nanocrystals easy to synthesize and handle can later block charge transport in a device or complicate compatibility with water and biology.

The easiest job to picture is keeping particles from sticking together. Two nanocrystals coated in long, flexible ligand chains cannot come arbitrarily close without compressing those chains, and compressing them costs energy. That steric crowding provides a barrier to aggregation. In some systems, ligands also carry charge, creating electrostatic repulsion. Either way, the nanocrystal becomes a colloid: a stable suspension of tiny solid objects in a liquid.

This stabilization is not cosmetic. It is what makes nanocrystal chemistry scalable and portable. If the product were a powder that immediately sintered or oxidized, it would be difficult to purify, characterize, and incorporate into real technologies. Colloidal stability turns the product into a liquid ink that can be pipetted, spun into films, mixed with polymers, or injected into a bloodstream *after* additional surface engineering. The magic trick of quantum dots in industry is not only that they emit narrowly in the visible; it is that they can be handled like a chemical reagent.

Yet the stabilization story understates the intimacy of ligands and the nanocrystal's electronic life. The surface is where the wavefunctions of electrons and holes feel the outside world most strongly. A broken bond can produce a trap state; a trap state can short-circuit luminescence. A ligand that binds well can satisfy, or at least neutralize, some of those problematic surface sites. The improvement can be dramatic: a sample can go from weakly glowing to brilliantly luminescent with what looks, in the lab notebook, like a small change in ligand choice or treatment.

It is tempting to describe this as "passivation," a word that suggests the surface is being pacified, rendered inert. But the surface rarely becomes inert. It becomes *regulated.* Ligands bind and unbind; different kinds of ligands compete; temperature and solvent change equilibria. In many systems, the ligand layer is better imagined as a dynamic marketplace than as a neatly tiled floor. This is not a defect of the field; it is a consequence of having a surface made of many atoms, each with slightly different environments, and a surrounding solution full of molecules that can interact with those atoms in multiple ways.

That dynamic character makes ligands powerful tools for controlling growth. When a nanocrystal grows, new atoms must attach to its surface. If the surface is densely covered in ligands that bind strongly, attachment becomes harder: incoming monomer must find a vacant site or displace a ligand. Growth slows. If ligands bind more weakly, attachment becomes easier and growth accelerates. If ligands prefer certain facets of the crystal—binding more tightly to one face than another—they can slow growth on those faces while leaving other faces relatively open, producing anisotropic growth and unusual shapes. A nanocrystal's final morphology is therefore not only a matter of crystal structure; it is the negotiated outcome of surface binding.

This helps clarify why coordinating solvents were so central in the classic hot-injection era. In many historically important syntheses, the "solvent" was not simply a heat-transfer medium. It was a ligand reservoir. Systems based on TOPO and phosphonic-acid-type ligands, for example, offered a chemically rich environment where binding, precursor reactivity, and growth rate were intertwined. The same molecules that kept particles dispersed also influenced how quickly precursors converted and how readily atoms could join the nanocrystal surface. When those molecules were changed, the entire tempo of the synthesis could change.

The optical consequence of all this is not merely that *better ligands make brighter dots.* Ligands can shift the balance between radiative recombination (light emission) and nonradiative recombination (energy lost as heat). Surface traps promote nonradiative pathways. A well-bound ligand can reduce the density of traps by satisfying dangling bonds or by altering the local electronic structure of surface atoms. A poorly matched ligand can do the opposite, leaving reactive sites exposed or even creating new electronic states through partial bonding and charge imbalance. If the emission spectrum is a census of size, the brightness is often a referendum on the surface.

There is also a social consequence: ligands decide where nanocrystals are allowed to go. A nanocrystal coated in long hydrocarbon chains will happily disperse in nonpolar solvents such as hexane or toluene, but it will avoid water as stubbornly as oil avoids a salad dressing without emulsifier. For display manufacturing, nonpolar dispersions can be convenient because they mix well with many polymers and can be processed into films. For biology, they are useless until the surface is transformed.

That transformation is usually accomplished by *ligand exchange*: replacing the original ligands from synthesis with new ones that confer different properties. The phrase sounds like a simple swap, but it is more accurate to treat it as chemistry in its own right, with pathways, byproducts, and surprises. James Owen and Jillian Park and coworkers (2008) drove this point home for CdSe nanocrystals by treating ligand exchange as a set of definable reactions rather than as "washing" or "coating." Using NMR and identification of organic byproducts, they showed that one can infer what ligands were actually bound and how they change during processing. This matters because a nanocrystal's surface is not a label you print on a bottle; it is a chemical interface where molecules rearrange and react.

The practical motivation for ligand exchange often begins with a frustrating observation. A synthesis yields beautiful nanocrystals in an organic solvent. The sample is bright, stable, and easy to store. Then someone tries to incorporate the dots into a device that requires electrical conduction, or into a water-based environment, and everything falls apart. The long organic ligands that stabilize the colloid act as insulating spacers between particles, blocking charge transport. Or they prevent the dots from dispersing in water. Or they desorb under heat or light, exposing surfaces and creating traps.

Exchanging ligands is an attempt to renegotiate these trade-offs. Shorter ligands can bring particles closer, improving charge transport in films. Polar ligands can make particles water-dispersible. Biomolecule-compatible ligands can allow attachment of antibodies or peptides. But each gain can come with a loss. Short ligands may bind less securely, increasing trap density and reducing brightness. Strongly charged ligands may improve water solubility but create new pathways for nonradiative recombination by altering local electric fields. Some exchanges can etch the nanocrystal surface, shrinking it and shifting its color. Others can destabilize the colloid, causing aggregation.

A useful mental picture is that a nanocrystal is never just a core of inorganic atoms. It is a hybrid object: an inorganic core plus an organic (or ionic) shell. The shell is not decorative; it participates in the nanocrystal's identity. When researchers discuss "the same dot" in two different environments—hexane versus water, or solution versus solid film—what they often mean is the same inorganic core with different shells. Those shells can change how the nanocrystal interacts with light, with charges, with its neighbors, and with living tissue.

This hybrid identity is why the surface has been a site of ongoing debate and refinement. It is tempting, especially in simplified cartoons, to draw a nanocrystal with a tidy monolayer of identical ligands arranged like bristles on a brush. The real picture is usually less clean. Nanocrystal surfaces often feature multiple binding motifs and dynamic equilibria rather than a single static "ligand monolayer" model. A ligand head group might bind in more than one way; some sites might be occupied by solvent molecules; some ligands might bridge two surface atoms; some might be protonated or deprotonated depending on the environment. Stoichiometry can be surprisingly hard to define: how many ligands are truly bound, how many are loosely associated, how many are simply trapped in the solvation layer?

These complexities are not just for specialists. They bear directly on why two batches made with "the same recipe" can behave differently. Slight differences in ligand purity, water content, or oxidation state can shift surface equilibria. A batch can have a similar size distribution yet differ in brightness because the surface trap landscape changed. A film can have the right color but poor stability because the ligand shell rearranges under illumination. When an engineer asks for "dots that survive 10,000 hours of blue pumping," they are asking, in part, for a surface that remains well regulated under stress.

The story becomes even more emotionally charged when nanocrystals are intended for medicine. A probe for imaging in living tissue must do more than glow. It must remain colloidally stable in salty, protein-rich fluids. It must avoid nonspecific sticking to cells. It must avoid triggering immune responses. It must remain bright despite oxygen, reactive species, and fluctuating pH. These demands are written in the language of ligands. The inorganic core may provide the color, but the ligand shell provides the biography: how the particle travels, where it accumulates, how long it persists, and how it is cleared.

The same is true, in a different register, for clean-energy applications and optoelectronics. A nanocrystal film used in a solar device or photodetector must transport charge. That means particles must be close enough that electrons or holes can move from one to the next, and the barrier between them is often the ligand layer. The ligand shell can be engineered to be thinner or more conductive, but doing so can expose surfaces and increase traps. This sets up a central tension in nanocrystal technology: the very molecules that make nanocrystals manufacturable can also limit their performance, and the art is to change the shell without ruining the core.

A small but telling laboratory ritual captures the ligand's power. After synthesis, nanocrystals are commonly purified by adding a "nonsolvent" that the ligand-coated particles dislike, causing them to precipitate, and then redispersing them in a clean solvent. This looks like simple cleaning, yet it can alter the ligand population, stripping weakly bound species and shifting equilibria. A sample can come out brighter or dimmer after purification, not because the core changed, but because the surface was edited. The chemist learns, sometimes painfully, that purification is itself a surface treatment.

Owen's 2008 work is a reminder that if one wants control, one must be able to *know* what is on the surface. NMR, especially when used thoughtfully, can separate signals from free ligands in solution and ligands bound to the nanocrystal, and it can identify reaction byproducts that reveal ligand transformations. That kind of detective work matters because nanocrystals are often discussed as if their surfaces are obvious: "they are capped with X." In reality, "X" can be a mixture, and the mixture can shift during synthesis, purification, storage, and use.

If there is a single dinner-party-level takeaway, it is that ligands are not packaging. They are the nanocrystal's interface with the world, and interfaces are where most failures occur. A nanocrystal that looks perfect in a vial may fail in a device because its ligands block charge flow. A nanocrystal that looks perfect in an organic solvent may fail in water because its ligands cannot keep it dispersed. A nanocrystal that looks perfect on day one may fail after weeks of illumination because its ligand shell slowly rearranges or desorbs, exposing trap sites that turn photons into heat.

Yet the same truth can be read optimistically. Ligands offer a lever arm that bulk semiconductors rarely provide. A silicon wafer's surface can be modified, but the wafer remains a macroscopic object. Nanocrystals are all surface and near-surface. By choosing, exchanging, and engineering ligands, chemists can turn one inorganic core into many functional objects: a printable ink, a stable backlight phosphor, a water-dispersible probe, a charge-transporting film. The light that emerges from the core is shaped, protected, and sometimes throttled by molecules that are small enough to hold in your mind as structures—yet influential enough to determine whether a nanocrystal remains a laboratory curiosity or becomes a reliable part of daily life.

5.4 Core-shell Architectures: Building a Better Energy Landscape

A bright nanocrystal is a small miracle of housekeeping. The crystal must confine an excited electron and the corresponding hole long enough for them to recombine by emitting a photon, rather than bleeding that energy away into heat through defects and surface mishaps. Ligands do a great deal of this housekeeping, but they have a limitation that becomes obvious once you think of the nanocrystal as an object with a hard boundary: ligands sit *outside* the semiconductor. They can bind, passivate, and regulate, yet the electronic excitation is still living uncomfortably close to the edge.

A core-shell nanocrystal changes the geometry of the problem. Instead of asking organic molecules to tame the outside of a semiconductor, chemists wrap the semiconductor in another semiconductor. The shell is chosen to have a wider band gap than the core, which is a way of saying it is energetically less hospitable for the excited electron and hole. The carriers are then encouraged to reside in the core, away from the outer surface where dangling bonds and chemical intrusions are most likely to create traps. If the core is the light emitter, the shell is both armor and architecture: it protects, and it reshapes the landscape in which electrons and holes move.

This is more than a materials upgrade. It is a shift in what the nanocrystal *is*. A core-only particle is a single material with a complicated surface. A core-shell particle is a heterostructure, the nanoscale cousin of layered semiconductors used in lasers and high-electron-mobility transistors. The difference is that, in a colloidal nanocrystal, the heterostructure is not a flat sandwich but a sphere-like object: a core wrapped in a shell, floating in solution, processed like an ink.

The basic motivation emerged quickly once high-quality colloidal nanocrystals existed. In the mid-1990s, researchers could already make CdSe nanocrystals that showed size-tunable color, but their brightness and stability were not always reliable, and their emission could be finicky under illumination. In 1996, Hines and Guyot-Sionnest reported strongly luminescing ZnS-capped CdSe nanocrystals, demonstrating that overcoating CdSe with a wide-bandgap ZnS shell can substantially increase luminescence, establishing core-shell passivation as a route to higher quantum yield and stability. That result had the feel of an inflection point: it suggested that the surface problem was not merely something to be patched with better ligands, but something that could be engineered inorganically.

A year later, Dabbousi and coworkers (1997) produced a systematic CdSe/ZnS core-shell size series, linking shell growth and structure to photoluminescence performance and emphasizing that shell growth can be locally epitaxial, with shell quality influencing emission. The phrase "locally epitaxial" carries an important idea in plain clothes: the shell is not just an amorphous crust. In good cases it grows with some registry to the core's lattice, like bricks laid following the pattern of the foundation. When that registry is decent, defects at the interface are fewer, carrier confinement is cleaner, and emission is brighter and more stable.

The concept that ties these improvements together is *band alignment*. In a semiconductor, the allowed energies for electrons and holes form bands separated by a gap. When two semiconductors touch, their bands do not line up automatically; they form an energetic step. If the shell has a wider band gap and the alignment is favorable, both the electron and hole prefer to sit in the core. This is the so-called type-I arrangement in broad terms: the core forms a potential well for both carriers, and the shell forms the barriers. That well-and-barrier picture is not a metaphor; it is how the wavefunctions distribute themselves. The more the carriers are confined to the core, the less they "feel" the outer surface, and the fewer chances there are for a surface defect to steal the excitation.

The same alignment idea also clarifies why shells can improve color purity and stability without changing the nominal core size. If an electron or hole wavefunction extends to the surface, it is sensitive to tiny changes: a ligand desorbing, an ion in solution, a bit of oxidation, a patch of disorder. Pulling the wavefunction inward makes emission less twitchy. The nanocrystal becomes, in a sense, more self-contained.

At this point it helps to put the energy story into a picture, because the mind naturally wants to know where the electron and hole "live" in a core-only particle versus a core-shell one.

Once this energetic picture is in place, the practical questions become more interesting. How thick should the shell be? How perfect must the interface be? Which material should be chosen as the shell?

Thickness is, at first glance, the simplest variable. A thicker shell increases the physical distance between the carriers and the outer surface. That distance can be protective in the same way that a thick winter coat protects you from wind: the wind still exists, but it does not reach your skin as easily. In nanocrystal terms, the carriers are less sensitive to whatever chemical drama is unfolding in the surrounding environment. Thick shells can also mechanically stabilize the particle and slow down processes that lead to degradation.

Thickness has another, more subtle consequence that many people encounter as a flicker in single-particle experiments: blinking. Some nanocrystals switch between bright and dark states under continuous illumination. The mechanisms are still a topic of detailed study, but surface charging and trap-mediated processes are central. A striking result from later work is that very thick (*giant*) shells can suppress blinking, suggesting that sufficiently thick inorganic shells can reduce sensitivity to the environment by distancing carriers from surface traps. In everyday terms, the particle becomes less interruptible. For applications where consistent brightness is essential, that kind of behavior is not a curiosity; it is a requirement.

And yet thickness is not a free lunch. As shells grow, they introduce strain if the core and shell lattices do not match well. Lattice mismatch strain, as already discussed, is the material's discomfort when two crystalline patterns of slightly different spacing are forced to fit together. At a small number of shell layers, the shell may accommodate the mismatch elastically. As thickness increases, strain energy accumulates, and the system looks for ways to release it. It can do so by forming dislocations and defects—exactly the kinds of imperfections that create trap states and undermine the optical gains the shell was meant to provide.

This is why the phrase "a wider-bandgap shell improves luminescence" is both true and incomplete. The shell can passivate the core surface and confine carriers, but it can also seed a new set of problems at the interface. Core/shell design is governed by band alignment, interface defects, and lattice mismatch/strain; the shell material must be chosen with all three in mind. A shell that is perfect energetically but poor structurally may disappoint. A shell that is structurally compatible but energetically leaky may also disappoint, confining one carrier well but letting the other wander.

The simplest success story, CdSe/ZnS, is also a good illustration of trade-offs. ZnS has a wide band gap, so it forms strong barriers that tend to keep carriers in CdSe, improving brightness and stability. But

ZnS also differs in lattice constant from CdSe, so thick, flawless ZnS shells are difficult. Many strategies in the broader literature can be understood as responses to that tension: use intermediate layers, use alloys, use graded shells that gradually change composition from core-like to shell-like, or select alternative shell materials that balance band alignment against strain. The details vary across material systems, but the design logic is consistent: manage energy *and* manage mismatch.

Shell growth also raises a question that sounds mundane but turns out to be profound: can a shell rescue a flawed core? If the core has a high density of internal defects, or if its surface is already rough and trap-rich, adding a shell may improve stability without delivering the hoped-for brightness. The shell can hide the core from the environment, but it cannot necessarily erase nonradiative pathways that are intrinsic to the core or embedded at a poorly formed interface. In that sense, the shell is not a magic eraser; it is a protective overcoat applied to whatever quality of object you started with.

This is where Dabbousi's emphasis on shell quality becomes more than a technical footnote. Shell growth can be locally epitaxial, but only if the conditions allow it. If shell precursors react too quickly, they may nucleate their own particles rather than coating the core. If the temperature is wrong or ligands are poorly matched, shell material may deposit unevenly, creating patches, pinholes, or roughness that act as new traps. The best core-shell particles often arise from shell-growth protocols that are deliberately patient: slow precursor addition, carefully chosen ligands, and temperatures that encourage atoms to find low-defect positions rather than freezing into disorder.

There is also a conceptual payoff to seeing core-shell particles as engineered energy landscapes rather than as "core plus coating." Once you grant yourself permission to engineer the landscape, you can design for behaviors beyond simple brightness. You can tune where the electron sits relative to the hole, which influences radiative rates and interactions with the environment. You can design multishell structures to balance confinement with strain management. You can, in some cases, create architectures where carriers are deliberately separated rather than co-localized, trading brightness for longer-lived charge separation useful in certain photochemical applications. The menu of possibilities is large because semiconductors offer many combinations of band gaps and alignments.

Still, the most humanly relatable benefit remains the simplest: reliability. A display backlight does not want a pigment that brightens and dims unpredictably. A biological assay does not want labels that bleach quickly or respond dramatically to small changes in salt concentration. Shells are a way of making nanoscale light sources behave more like dependable materials and less like temperamental surfaces.

The price of that dependability is complexity. A core-shell nanocrystal asks the chemist to control not only size and surface ligands, but interfaces and strain. It asks the engineer to remember that a particle can be chemically stable yet electronically flawed; mechanically intact yet optically noisy. It also adds a new kind of vulnerability: shells modify, but do not eliminate, the importance of interfaces and surface-related electronic states. A core-shell particle still has an outer surface, and it still has an interface buried inside. It has, in other words, two places where imperfections can concentrate.

When core, interface, and shell come together well, the result can feel almost uncanny in its steadiness: a nanocrystal that keeps emitting with high efficiency, resists chemical insults, and maintains its color under prolonged illumination. That steadiness is not an accident of nature. It is a built landscape, atom by atom, designed so that an excited electron and hole prefer the calm interior of the core over the noisy boundary of the world outside.

Chapter 6

Surviving the Real World: Heat, Oxygen, and the Long Goodbye

The hardest part of making quantum dots useful is not persuading them to glow once, but persuading them to keep glowing after months of heat, light, air, and neglect. What looks like a tiny materials problem becomes a story about leaky barriers, imperfect surfaces, accelerated lifetimes, and the uncomfortable fact that at commercial scale, consistency is itself a form of durability.

6.1 When Brightness Learns to Fade

A new display, seen in a darkened room, can feel almost indecently vivid. Reds look lacquered; greens look freshly cut; whites have that clean, icy snap that makes paper seem warmer by comparison. When a color-conversion layer is doing its job well, the physics stays politely invisible. The little crystals absorb light, then return it in narrower, better-aimed colors, and the eye simply reports: sharper, brighter, truer.

What happens in the real world is rarely a sudden failure. The colors do not usually vanish in an instant, like a bulb popping. Instead the performance drifts. A small dimming that the owner cannot quite swear to, then a larger one. A white point that slides warmer or cooler. The clean edges of a color that blur as the emission line broadens. Sometimes the device still lights, still looks "fine," but the efficiency has quietly worsened: more electricity, more backlight power, more heat, to get the same perceived brightness. Degradation is not one thing; it is a family of slow insults that all share a theme. They find their way to the nanocrystal's surface.

That surface is where the beauty and the fragility live side by side. In a macroscopic crystal, most atoms sit comfortably inside the solid, surrounded by neighbors in a tidy repeating pattern. At the nanoscale, a large fraction of atoms sit at or near the outside, where the pattern ends. Bonds are "unfinished," and chemistry that would be irrelevant in a bulk chunk becomes decisive. The electronic consequences of that unfinished surface can be described in a single, practical phrase: traps. A trap is a place where an excited electron or the corresponding "hole" (the absence it leaves behind) can get caught and lose its energy without giving off light. The dot still absorbs energy, but the return on that energy changes from photons to heat.

The everyday scorekeeper for this is the quantum yield: the fraction of excitations that end as emitted photons. A quantum yield near 1 means nearly every absorbed photon comes back out as light. A quantum yield sliding down toward 0 means energy is leaking into non-radiative pathways. The slow tragedy of aging is often a slow rewriting of that ledger. The radiative route is still available, but the non-radiative alternatives become easier to reach.

It helps to picture the dot not as a jewel but as a small town with roads. The ideal case has one broad, well-maintained highway from excitation to emission. Degradation is the quiet construction of side roads that lead to dead ends: surface states, chemical byproducts, tiny structural rearrangements. The emission does not need to be blocked completely; it only needs competition. If, after excitation, there are now several ways to lose energy without emitting, brightness drops even if the dot remains intact and recognizable.

Four outside pressures do much of the construction work: oxygen, water, heat, and light. They are ordinary, unavoidable features of the world where screens glow, lamps run, and imaging probes are asked to behave inside living tissue.

Start with oxygen, because it is both chemically aggressive and culturally invisible. Oxygen feels gentle; it is what we breathe. But for many semiconductors it is a hungry reactant, especially when light is involved. A nanocrystal's surface offers oxygen both access and leverage: access because the surface is most of the object, and leverage because the surface is already undercoordinated and often stabilized only by a thin molecular layer.

One of the most revealing demonstrations of oxygen's role is also one of the simplest. Careful studies of how CdSe nanocrystals change under everyday storage-like conditions reveal a clear pattern: light plus oxygen accelerates the loss and alteration of photoluminescence, while darkness and oxygen-free environments preserve the optical properties much longer. That sounds almost too commonsense to be worth publishing, until you remember how much of nanoscience happens under bright lab lights, on benchtops, in vials that feel "closed" because a cap is screwed on. A slow leak of oxygen through imperfect seals, or a tiny headspace of air, becomes part of the experiment whether you want it or not. The combination of oxygen and illumination is a particularly efficient recipe because light does not merely reveal the damage; it helps drive it. Excited carriers can participate in surface reactions, and reactive oxygen species can be generated that attack the surface more aggressively than oxygen in its ground state.

Chemically, surface oxidation can do two kinds of harm at once. It can directly create new defect sites: local patches where the composition is no longer the tidy semiconductor lattice but an oxide or a mixed phase. It can also undermine passivation by changing how ligands bind. Either way, the electronic result is similar: more traps. Those traps shorten the excited-state lifetime and redirect energy away from emission.

Water is a close partner to oxygen in practice, even if it plays different roles depending on the dot composition and the surrounding materials. In many real devices, humidity is not a dramatic flood; it is a slow diffusion of moisture through polymers, adhesives, and microscopic seams. Water can help oxygen reach reactive sites by swelling a host matrix or by altering ligand packing. It can participate in hydrolysis reactions that weaken ligand attachment or change the surface composition. And in layered devices, water can attack not just the emitter but neighboring functional layers, changing the local environment in ways that feed back into the emitter's behavior.

Even without any oxygen at all, heat alone can age a nanocrystal. Heat is not a chemical species, but it is a powerful permission slip. Atoms vibrate more; diffusion becomes easier; weakly bound ligands detach and fail to reattach; interfaces slowly rearrange. In a perfect bulk crystal, a modest temperature rise may do little besides broaden spectral lines temporarily. In a nanoscale object, modest temperature can push it over small energetic barriers that separate one stable shape from another. If a nanocrystal slowly rounds off sharp facets, or if atoms migrate in a way that slightly changes its effective size or composition profile, the band gap shifts. A shift in band gap is, to the eye, a change in color. This is one reason "color shift" is such a sensitive failure mode: you do not need massive chemical destruction to change the emission wavelength. A small physical change, multiplied by the exquisitely size-dependent nature of confinement, can move the color point enough to be noticed in a calibrated display.

Heat also interacts with the dot's surroundings. In a film, nanocrystals are not isolated; they are crowded, pressed into a matrix, sometimes close enough for energy transfer between neighbors. A polymer host that is stable at room temperature might soften slightly at elevated temperature, allowing particles to move and cluster. Clustering can change optical behavior even if the particles themselves are chemically unchanged, because the local electromagnetic environment changes and because energy can hop to less emissive sites. The observer sees a dimmer, less saturated output and may never suspect that part of the story is mechanical.

Light, the very thing that makes these materials useful, is another stressor. Photobleaching is the name borrowed from organic dyes: a gradual loss of fluorescence under illumination. In nanocrystals, photobleaching often reflects a mixture of chemical and electronic changes. Under sustained excitation, the dot is repeatedly pushed into excited states. That repetition increases the chances of rare events: a charge escaping to the surface, a bond breaking, a ligand being knocked loose, a reactive intermediate forming when an excited carrier meets oxygen or water. Photons become tiny hammers tapping on the same place. Most taps do nothing; one tap out of a million triggers a change. Given enough time, rare events become inevitable.

Some photodamage is not purely chemical but electrostatic. A dot can become charged during operation, especially in electroluminescent devices where carriers are injected, but even under optical pumping where charges can be separated by local fields or by interfacial processes. A charged dot often emits poorly because a fast nonradiative process becomes available: Auger recombination. In Auger recombination, the energy that might have produced a photon is instead given to a third carrier, which then thermalizes. The dot warms by an imperceptible amount, but the photon never appears. The practical symptom can resemble ordinary dimming, but the mechanism is distinct: it is not that the band gap has changed, but that radiative recombination is being outcompeted by a faster, non-radiative channel that opens when extra charge is present. This is one bridge between the microscopic world of traps and the macroscopic world of "my screen seems duller": if charging events become more frequent as surfaces degrade, then the periods of quenched emission become more common, and average brightness drops.

There is a subtler way light can erode color quality even when brightness seems acceptable. The clean color that makes these materials so valuable comes from a narrow emission spectrum. When a dot is pristine and well passivated, most excitations recombine through the same well-defined route, producing photons clustered around the same energy. When traps and defects proliferate, recombination can occur through multiple routes with slightly different energies. The spectrum broadens. The color becomes less pure. In display language, the "gamut" shrinks not only because the peak wavelength drifts, but because each primary becomes a little smeared.

This slow broadening has an emotional analogue that anyone who has owned a beloved garment recognizes. The garment does not tear in one catastrophic snag. It pills. The color fades in the wash. The seams loosen. Each change is small enough to ignore, but together they alter the experience. The physics version is that a dot can remain a dot while becoming a worse emitter. The nanocrystal is still there; it still absorbs; it still emits; it simply does it with less discipline.

The surface is the arena where most of these changes begin, and it is worth lingering on why. A dot is often synthesized with organic ligands: molecules that bind to surface atoms and satisfy their dangling bonds, preventing the formation of trap-like surface states and keeping particles dispersed in solution. These ligands are not painted on like varnish; they are in dynamic equilibrium. They can detach and reattach. They can be displaced by other molecules. They can slowly desorb when heated. Under illumination, they can be chemically transformed. If the ligand coverage thins unevenly, certain patches of surface become exposed. Those exposed patches are invitations to oxygen and water and also sites where carriers can localize. That localization can itself accelerate chemistry: a carrier sitting near a bond can make that bond easier to break. A defect can therefore be both a cause and a consequence of chemical change, a feedback loop whose outward sign is creeping loss of quantum yield.

A concrete example makes the chain of cause and effect less abstract. Imagine a color-conversion film in the backlight of an LCD television. It lives in a warm compartment, near LEDs, inside a carefully engineered stack that still must tolerate years of being on for hours at a time. When the TV is new, the film converts part of the blue LED light into red and green with high efficiency and narrow spectral lines. Over time, a tiny amount of oxygen that permeates through barrier layers reaches the dots. Under constant blue excitation, photooxidation at weak points on the surface forms new trap states. Each trap increases the odds that an excitation will die without emission. The film's conversion efficiency drops. To keep the picture bright, the system may drive the backlight a little harder, raising temperature and accelerating diffusion. The user sees a slightly dimmer picture, or a white balance that no longer matches what it did out of the box. If the spectrum broadens, colors feel a little less crisp, a change that is hard to describe but easy to sense when compared side-by-side with a new device.

The frustrating part for engineers is that this story can unfold without any single dramatic defect that can be easily "fixed." There is no one crack to fill, no single chemical reaction to eliminate. There are pathways, and the job is to make all the bad pathways unlikely enough that a product survives its warranty and, ideally, much longer.

Because the mechanism is distributed, the symptoms can be confusing. Dimming could come from fewer photons being emitted per absorbed photon (a quantum yield drop), from fewer photons escaping the device (because a matrix yellows or scatters), from a shift in the absorption and emission balance, or from changes in the surrounding layers that alter how energy reaches the emitter. Color shift could come from slight changes in nanocrystal size or composition, from differential aging of different color components, or from changes in reabsorption within a film. Broadening could signal rising defect-mediated emission, but it can also reflect increasing heterogeneity as the population becomes more uneven: some particles degrade faster, so the ensemble emission becomes a mixture of "healthy" and "injured" emitters.

All of this makes reliability feel, at first, like a disappointment: why should something so precisely engineered be so vulnerable to air and warmth and light? But there is another way to read the same facts. The sensitivity is a consequence of what makes these materials powerful. When the active region of an emitter is measured in nanometers, the boundary is never a minor detail. The boundary *is* the device. The same confinement that turns size into color also turns small chemical changes into macroscopic optical differences.

A useful mental shift is to stop imagining an emitter as a static bead of solid material. It is closer to a living interface between a crystal and its environment, continually negotiating what counts as "inside" and what counts as "outside." When that negotiation goes well, the emitter looks like a simple source of pure color. When it goes badly, the world's most ordinary molecules—oxygen and water—rewrite the rules, and the emitter gradually becomes what bulk chemistry would predict: a reactive surface that prefers to turn into something else.

In the end, fading is not the opposite of brilliance; it is the cost of hosting brilliance in a place where chemistry can reach it. The world is full of photons, heat, and air, and at the nanoscale those ordinary facts can be as decisive as any clever design.

6.2 A Coat Against the World

A nanocrystal that performs beautifully in a vial is a bit like a soap bubble that holds its shape in still air: impressive, and not yet proven. Once it is asked to work inside a display stack, a lamp, or a biological assay, the world starts touching it from every side. Oxygen and moisture diffuse. Heat cycles with use. Light arrives not as a gentle probe but as a constant stream. The surprise, for newcomers, is that the glowing crystal in a commercial product is rarely a bare object. What actually gets engineered is a protected system: a core that does the emitting, wrapped in multiple layers that exist mostly to fend off everything else.

A useful way to think about this protection is *defense in depth.* No single barrier is perfect. Each layer addresses a different vulnerability, and each layer has its own failure mode. The same design choices that keep an emitter bright can also complicate manufacturing, change how the particles pack into films, or make it harder for electrical charges to reach the emitter in devices that rely on injection. Protection is not a cosmetic afterthought; it is a second, intertwined problem that can dominate whether a technology makes it out of the laboratory.

The innermost layer, the ligand coat, is the one most people never picture. Ligands are organic molecules attached to the nanocrystal surface. They have two jobs that are easy to describe and difficult to optimize at once. First, they passivate: they bind to surface atoms and reduce the number of dangling bonds that would act as traps. Second, they provide compatibility: they make the particles soluble in a chosen solvent and later help them mix into a polymer or other host without clumping.

A ligand coat is not a hard shell. It is more like a crowd around a celebrity: dense in places, thinner in others, and constantly shifting. Heat can loosen the crowd. Competing molecules can elbow their way in. Light can change the chemistry of the ligands themselves. That makes ligands necessary but not sufficient, which is why modern emitters are so often built as core–shell structures. As discussed earlier, inorganic shell passivation is one of the major reasons these materials became practical: a wide-bandgap semiconductor shell provides a sturdier, more coherent boundary than a layer of organic molecules alone.

The shell does something psychologically comforting: it looks like protection you can draw. But it is more than a physical barrier. It is also an *energetic* barrier. Carriers generated in the core have a harder time reaching the surface. That separation matters, because many of the worst fates begin at the surface: trapping, chemical reactions, charging events that trigger fast non-radiative pathways. Move the action inward, and the surface becomes less consequential.

The most striking demonstration of how far this can be pushed came from so-called *giant* nanocrystals. In work associated with Vladimir Klimov's group (2010), very thick inorganic shells around CdSe cores suppressed Auger recombination dramatically. Auger recombination, as noted earlier, is the process that turns what could have been light into heat when extra charge is present. A thick, wide-gap shell changes the spatial distribution of carriers and reduces the efficiency of that energy-dumping process. In practical terms, the emitter becomes more resilient against charge-induced quenching. It is harder to force it into long dark periods, and its brightness becomes less fickle under conditions that would otherwise promote charging.

The appeal of this result is that it reframes stability as designable. If a shell can suppress Auger processes by orders of magnitude, then the common complaint—that nanoscale emitters are doomed to flicker and fade—starts to sound less like a law of nature and more like an engineering challenge. Around the same time, Mahler and colleagues (2010) showed that blinking suppression depends strongly on shell thickness, and that composition and interfacial structure also play a major role. The implication is subtle but crucial: making a shell thicker is not a magic dial you turn to "more stable." The quality of the buried interface, defects in the shell, and whether the transition from core to shell is abrupt or graded can decide whether the coat is watertight or riddled with microscopic leaks.

Those leaks are not just places where oxygen can sneak in. They are also places where electronic traps can form. A defect in the shell can create a local energy pocket that captures a carrier. Once a carrier is localized, chemistry gets easier: charges and excited states can drive reactions that would be unlikely in a uniformly passivated structure. This is why photostability is not guaranteed by thickness alone. A thick coat with weak seams may still fail under sustained illumination, and the failure can look frustratingly similar to the failure of a much thinner shell: a gradual loss of quantum yield, a drift in spectrum, an increase in non-radiative decay.

The protection story continues outward. A core–shell particle with ligands is still a nanoscale object with an exposed surface, now the surface of the shell. To turn a dispersion of particles into something a manufacturer can handle—a film in a backlight unit, a downconversion layer in lighting, a bead in a biological assay—the particles must be embedded in a host matrix. Polymers are popular hosts because they are processable: they can be cast, coated, cured, and laminated. Glasses and inorganic matrices are attractive because they are often better barriers against oxygen and water and can be more stable under heat and light, but they can be harder to process at low temperature and may introduce their own compatibility problems.

Silica occupies an interesting middle ground. It can behave like a glassy, inorganic barrier at relatively gentle processing conditions if it is built around the particles in solution. Gao and colleagues (2010) demonstrated a dual-layer encapsulation approach combining silica and polymer, aiming for "remarkable stability." The phrase is not marketing fluff; it reflects a genuine logic. A polymer layer can provide mechanical compatibility and flexibility, while silica provides a tougher chemical shield. If oxygen and water must pass through multiple, chemically different obstacles, their diffusion slows. That slowing buys time.

Silica embedding is also attractive because it can protect not only brightness but chromaticity: the stability of the actual color point. For display makers, chromaticity drift is a particularly cruel failure mode. Human vision is forgiving of some dimming; it is less forgiving of a white that slowly shifts. Work on green-emitting InP/ZnS core–shell particles embedded in silica (2018) emphasized that silica can help preserve color coordinates by limiting oxygen contact. This matters because oxidation and surface chemistry changes can affect different parts of an ensemble differently, broadening the spectrum and nudging the peak. When the host matrix itself is a decent barrier, the dots are not asked to do all the defensive work alone.

At the outermost scale sits packaging: barrier films, laminates, encapsulants, and the seals that keep ambient air away from the active layer. This is where the nanoscopic and the everyday meet most directly. A consumer never thinks about the water vapor transmission rate of a film, yet that number can decide whether a device looks the same after years on a living-room wall. Packaging is also where cost and manufacturability become non-negotiable. An exotic barrier that performs wonderfully but requires slow, expensive deposition may be unacceptable. Conversely, a cheap barrier that leaks slowly might still be fine if the inner layers are robust enough. Reliability becomes a budget that can be spent in different places.

The temptation is to imagine these layers as independent: ligands do their job, shells do theirs, hosts do theirs, packaging does theirs. Real systems are messier because interfaces talk to each other. A ligand that stabilizes a colloid in solution might be a poor match for a polymer matrix, leading to aggregation during film formation. Aggregation can change optical behavior through reabsorption and energy transfer, and it can also create local hot spots under illumination. A shell that is chemically robust might have a lattice mismatch with the core that creates strain, which can introduce defects that compromise photostability. A host polymer might contain residual catalysts or impurities that slowly attack ligands. Packaging adhesives can outgas small molecules that diffuse inward and displace surface-bound ligands. Each step outward introduces new chemistry.

Protection can also collide with performance in a more direct way: the same layers that keep oxygen out can keep electrical charges out. In color-conversion applications, where the emitter is optically pumped, that is not a fundamental problem. In electroluminescent devices, where carriers must be injected into the emitter layer, it becomes central. Long organic ligands are wonderful for keeping particles bright and separated; they are terrible at letting charges move. Shorter ligands can improve charge transport but may leave the surface less protected. Some of the most frustrating gaps between photoluminescence performance (bright under optical excitation) and electroluminescence performance (bright under electrical injection) trace back to this tension. Under injection, ligands can also face electrochemical stress. Shen and colleagues (2020) argued that ligand electrochemical instability can limit operational lifetime, and that choosing more electrochemically stable ligands helps bridge that gap. The point is not that oxygen and water stop mattering in devices; it is that protection must anticipate the specific assaults of operation, including electric fields and injected charges.

There is a historical rhythm to how this layered approach emerged. Early colloidal nanocrystals were often treated like chemical curiosities: gorgeous spectra in a cuvette, delicate handling, rapid degradation if you were careless with air or light. As the field moved toward products, chemists began to treat stability as part of the optical design itself. Thick shells, graded interfaces, improved ligands, better encapsulation, and better barrier films were not separate inventions; they were successive answers to the same question asked at different scales: how do you keep the emissive core in the conditions where it behaves like the idealized object from the lab?

A concrete example makes the layering feel less abstract. Consider a backlight film used to sharpen the colors of an LCD. The emitter particles are embedded in a polymer sheet. That sheet sits near bright blue LEDs, under heat, for thousands of hours. The polymer is chosen for optical clarity and manufacturability, but it is not a perfect oxygen barrier. A manufacturer may add an external barrier film to slow oxygen and moisture ingress. Inside, the particles may have a core engineered for the desired emission wavelength, a shell chosen to isolate the core and reduce non-radiative losses, and ligands chosen to keep the particles dispersed and compatible with the polymer. If the shell is too thin or defect-rich, oxygen that arrives slowly will still find reactive sites. If ligands detach under heat, the shell surface becomes chemically vulnerable. If the polymer matrix allows particles to move and cluster over time, optical properties can drift. The final product's stability is therefore not the property of a material in isolation; it is the property of a *stack* in which each layer can compensate, partially, for the weaknesses of others.

This layered mindset also reshapes how “failure” is interpreted. A dimming screen is not automatically telling you that the core material is intrinsically unstable. It might be telling you that the barrier film chosen for cost reasons leaks enough oxygen over years to slowly oxidize surfaces. A color shift might be telling you that the host matrix permits subtle rearrangements, or that a subset of particles with thinner shells are degrading first. A product is the average of many particles and many interfaces; protection is an attempt to make that average behave predictably.

The final, slightly humbling lesson is that a coat can change the very thing it protects. Add too much inorganic shell and you may shift absorption or change how energy flows through a film. Wrap particles too thoroughly and you may make them optically safe but electrically isolated. Choose a host matrix with superb barrier properties and you may introduce optical scattering or processing difficulties. Yet when protection is well matched to the application, the result can feel almost magical: a material built from nanoscopic crystals, each one with a fragile surface, performing for years in warm air under relentless light, as if chemistry had agreed to take a holiday.

6.3 Borrowing Time from Heat and Light

A new material has a public-life problem. If it is meant for a television, a surgical imaging probe, or a room light, it will be judged on the scale of years. Yet the people developing it live on the scale of weeks and quarters. Nobody wants to discover, after shipping a million devices, that a subtle chemical pathway takes eighteen months to show itself. Reliability science exists to settle that mismatch by a slightly audacious strategy: make time run faster.

The basic bargain is familiar from cooking. A stew that takes hours at a low simmer can be pushed along by raising the heat, but too much heat gives you a different stew. Accelerated aging tries to find the region where "faster" still means "the same mechanism," just hurried. It is never completely comfortable, because the moment you push a material harder than it will be pushed in real use, you risk opening up a degradation route that would never have mattered. That is why lifetime claims for nanoscale emitters are not just numbers; they are arguments about causality.

The most common accelerants are the simplest: higher temperature, higher humidity, and higher light intensity. Temperature is the blunt instrument. Many unwanted reactions become dramatically faster with heat, because more molecules have enough energy to climb over the barrier separating "stable" from "reacted." A familiar rule of thumb says that reaction rates often rise strongly with temperature, and engineers often lean on Arrhenius-style reasoning to quantify that rise. You do not need the full formula to appreciate the practical point: raising temperature by a few tens of degrees can compress months into days. It is the reason a device might be held at 85°C for hundreds of hours, not because anyone expects it to run at that temperature in a living room, but because the chemistry that would otherwise unfold slowly can be forced to reveal itself.

Humidity is more selective. It targets pathways where water participates directly or where it acts as a facilitator: swelling a polymer, changing ionic mobility, or altering how protective molecules sit on a surface. A classic damp-heat exposure, such as 85°C at 85% relative humidity, is the sort of test that makes packaging engineers nervous and materials chemists curious. It asks a question that normal operation often asks more politely: how much does your system rely on staying dry?

Light is the trickiest stressor, because it is both the working input and a chemical trigger. The same photons that create the desired emission can also create excited carriers energetic enough to participate in unwanted reactions. Photooxidation is a prime example: illumination in the presence of oxygen can accelerate chemical changes at nanocrystal surfaces, which in turn alter exciton recombination and photoluminescence stability. This is not merely "light causes heating." It is light changing the reaction landscape. A shelf-stability test that exposes a sample to intense blue light in air can be brutally informative, but it can also be misleading if the device in the field would have excellent oxygen barriers and would never see that combination of oxygen and photon flux.

The relationship among these stressors often turns out to be more revealing than any single one. Consider a color-conversion film used in an LCD backlight. In normal use it may sit at a moderate temperature, bathed in blue light, sealed behind barrier layers designed to keep oxygen and moisture out. In accelerated testing, one might increase the blue intensity, raise the temperature, and deliberately challenge the barrier with humidity. Each change makes failure quicker. Yet failure under combined stress may not be the same as failure under typical use. A polymer matrix that survives light at room temperature may soften at higher temperature, allowing nanoparticles to rearrange and aggregate. A barrier film that is adequate at moderate humidity may allow moisture ingress at elevated humidity that never happens in a living room. The accelerated test produces real failure data, but it may be diagnosing a vulnerability that is irrelevant in practice, while a different vulnerability remains hidden.

A concrete example from the laboratory captures this tension between speed and relevance. Studies of how CdSe nanocrystals change under conditions that are not exotic at all—exposure to oxygen and light during storage—make the point. The message was unglamorous and therefore powerful. Brightness and spectral properties could drift significantly faster when light and oxygen were present together; darker, oxygen-free storage preserved properties far longer. That kind of result is a reliability lesson disguised as a chemistry lesson: your "control" conditions may already be an accelerated test if you leave samples under bright lights in air. It also hints at a deeper point about mechanism. Oxygen alone may be slow to react with a well-protected surface, but an excited state can tip the balance, creating reactive intermediates or local charge distributions that make oxidation easier. Acceleration is sometimes a matter of *turning on* a pathway, not merely speeding it up.

Because the goal is prediction, not just breakage, reliability tests need metrics that connect to what people care about. For light emitters those metrics often start with brightness retention: how long until output falls to some fraction of its initial value, perhaps 90% or 50%. Color stability matters too: the drift of the emission peak or the shift of chromaticity coordinates. In displays, a slight change in a primary color can be more damaging than a slight dimming, because it changes skin tones, whites, and brand colors in ways viewers detect instinctively. For bioimaging, the worry can be different: photobleaching during an experiment, where losing signal after minutes rather than years is the failure. The same material can therefore have multiple "lifetimes," depending on the job.

A subtle but important choice hides inside any accelerated protocol: are you testing the emitter as a material, or the device as a system? A film in a vial can be illuminated and heated to test photostability and thermal stability of the emitter in isolation. A finished electroluminescent device can be driven at high current to test operational lifetime. These are related but not interchangeable, because devices add new stresses: electric fields, injected charges, joule heating, and a crowded stack of interfaces that can degrade in their own ways.

This difference is particularly stark for QLEDs, where operational degradation can proceed via thermally and electrically driven mechanisms even if the emitter itself looks robust in photoluminescence tests. Encapsulation reduces oxygen and moisture ingress, but it does not eliminate heating under high brightness, nor does it prevent chemical changes driven by injected carriers. In other words, a perfect oxygen barrier does not guarantee a long operational lifetime. Heat management and device engineering remain necessary because some failure routes are internal to the electrical operation: changes in transport layers, interfacial reactions, charge imbalance that promotes charging events, and local hotspots that accelerate reactions in specific regions of the device.

Ligands, which in optical tests behave mainly as passivating and compatibility molecules, acquire a new personality under injection. Ligand chemistry can be an operational lifetime limiter: under carrier injection, ligands may undergo electrochemical reactions. The phrase sounds technical, but the image is simple. If a molecule is sitting at the surface of an emitter and you repeatedly push electrons and holes through the region, you are not merely exciting the crystal; you are running a microscopic electrochemical cell. Some ligands are happy to sit through that. Others are not. Shen and colleagues (2020) demonstrated that choosing more electrochemically stable ligands can improve operational lifetime and help close the gap between photoluminescence performance and electroluminescence performance. That is an important reliability moral: a material can be chemically stable in air and still be unstable under the particular electrochemical stresses of a device.

Accelerated testing for devices therefore often raises current density or brightness to speed the processes that dominate in use. Again, there is a trap. Pushing a device to very high current may heat it into a regime it would not normally reach, causing a thermal failure that dominates the data. Or it may create charge densities that promote mechanisms absent at normal brightness. The test becomes a stress torture rather than a time-compressor. Yet testing at normal brightness may take too long. Engineers thread this needle by using multiple stress levels and checking whether the failure signature remains the same.

Failure signatures are where modern reliability work has become more sophisticated, and more honest. “It got dimmer” is a symptom, not a diagnosis. Researchers increasingly use operando and nondestructive approaches to identify which channel dominates under given operating stresses. The word *operando* simply means “while operating”: measuring as the device runs, rather than after it has failed. Nondestructive means you can keep measuring the same specimen over time rather than sacrificing it for post-mortem analysis at each stage. These approaches try to answer questions that accelerated tests alone cannot: Is the emission drop tied to growing nonradiative recombination in the emitter, to reduced charge injection, to interfacial energy-level shifts, to chemical changes in a transport layer, or to a barrier breakdown allowing oxygen in? Without that mechanistic attribution, extrapolating from accelerated data is guesswork with nicer plots.

There is a revealing analogy to medicine. A fever tells you something is wrong; it does not tell you whether the cause is viral, bacterial, autoimmune, or heatstroke. Treating the fever can buy time, but it might also mask the disease. Accelerated tests can likewise show that a system is vulnerable without telling you why. Better diagnostics are the equivalent of identifying the pathogen.

A newer idea in the reliability toolbox makes this explicit: intentional pre-aging, sometimes called conditioning. Recent (2026) work introduces solution-phase pre-aging/conditioning strategies for InP QLED emissive layers to decouple intrinsic dot-layer degradation from full device-stack effects. The motivation is both practical and philosophical. Many devices show "early-life" drift: a rapid change in the first hours or days, followed by a slower, steadier decline. If you are trying to compare materials, that early drift can dominate and confuse, because it may reflect reorganization of ligands, settling of interfaces, or the consumption of reactive impurities rather than the long-term wear-out mechanism. Conditioning aims to let the system get past its initial rearrangements under controlled conditions, so that subsequent tests probe the slower, more relevant processes.

This is not cheating; it is acknowledging that materials often have a youth. A new leather shoe changes shape in the first week and then lasts for years. If you want to know whether the shoe will last, you may care less about the first-day stiffness than about the long-term cracking. Conditioning is an attempt to separate the break-in from the breakdown. For InP-based emitters, which are attractive partly because of reduced regulatory pressure compared with cadmium-based materials, this kind of strategy reflects a broader maturity in the field: stability is being treated as a time-dependent trajectory that can be shaped, not just a fixed property to be measured once.

All of this brings us back to the core problem: extrapolation is not prediction unless the mechanism stays the same. Arrhenius-style models can be powerful when the same reaction controls the rate across the temperature range. They can be dangerously confident when a different pathway takes over at higher temperature or higher humidity. A photochemical oxidation pathway can dominate under bright light in air; a ligand desorption and aggregation pathway can dominate under heat; an electrochemical ligand breakdown pathway can dominate under injection. Each pathway has its own dependence on stress, and sometimes the combined stresses create entirely new behavior. A model that fits the accelerated data beautifully can still be wrong about real use.

Reliability work therefore tends to look less like a single grand test and more like a conversation among tests. Thermal storage asks about heat-driven restructuring and diffusion. Damp-heat asks about moisture ingress and hydrolysis-related problems. Photostability tests ask about photooxidation and trap growth. Device stress tests ask about electrical and thermal operation. If several tests point toward the same microscopic change, confidence rises. If different tests produce different signatures, that is not failure; it is information. It tells you the system has multiple vulnerabilities, and which one you need to address for a given application.

There is also a human side to accelerated aging that is easy to miss. Teams building new emitters are often under pressure to show progress quickly. A short, harsh test can feel like a verdict: pass or fail. But reliability is rarely binary. A material that fails under extreme humidity might still be excellent in a well-sealed display. A material that ages under intense illumination in air might be perfectly adequate in a device with robust oxygen barriers. The art is to choose tests that match the intended life, and to interpret failures as clues rather than disqualifications.

In the end, the most valuable thing an accelerated test can do is not to produce a single lifetime number, but to reveal which part of the protected system is negotiating with the world and losing. Once that culprit is named—a photooxidation route that needs better oxygen blocking, a thermally activated diffusion that needs better heat spreading, an electrochemical ligand weakness that needs a different molecular choice—time stops being a purely external enemy. It becomes a design parameter, something that can be budgeted, traded, and, with enough care, borrowed.

6.4 The Hidden Reliability of Sameness

A manufacturer does not ship a single nanocrystal. They ship a population: trillions of emitters spread across a film, a pixel array, a vial of imaging reagent, or a roll of laminated backlight material. In the lab, it is easy to fall in love with a best-case specimen: the sharp spectrum, the high photoluminescence quantum yield, the tidy microscopy image that flatters the synthesis. In products, the weak links get equal voting rights. A device can be built from emitters that are, on average, excellent and still age badly if a minority of them begin life slightly compromised.

This is the hidden reliability of sameness: uniformity is not only about meeting a color specification on day one. It shapes how the system *fails*. A tight distribution of sizes, shell thicknesses, ligand coverage, and defect densities yields not only cleaner color but more predictable aging. A broad distribution does something crueler than mere imperfection: it creates a built-in "early failure" subpopulation that will darken first, drift first, and seed larger changes around it.

The logic is familiar in other settings. A chain is not "mostly strong"; it is as strong as its weakest link. A choir is not "mostly in tune"; the sour notes dominate the listener's impression. In emissive films and devices, the analogous issue is that dim or unstable particles do more than contribute less light. They can also absorb light without returning it efficiently, effectively acting as tiny heaters. They can accept energy from their brighter neighbors through energy transfer and then waste it non-radiatively. They can trap charges, creating local electric fields that promote further charging and quenching nearby. Variability is not a passive statistic; it can create feedback.

The most obvious face of variability is size distribution, because confinement ties size to color. A population with a narrow size spread emits in a narrow band. Broaden the size spread and the emission line widens. That widening is not a mere aesthetic penalty. In displays and lighting, broader emission means poorer color purity, which can reduce gamut and make it harder to hit standardized color points. It also turns aging into a more visible story. If different sizes age at different rates (or if they shift in wavelength differently under heat or oxidation), the color point can drift in time even when the average brightness seems acceptable.

Shells introduce another axis of sameness that is easy to underestimate. Two emitters can have identical core sizes and therefore identical initial emission peaks, yet differ in long-term stability because one has a slightly thinner shell or a shell with a few more defects. In earlier discussions, thick inorganic shells were described as chemical armor and an energetic barrier. That armor is only as reliable as its consistency. If most particles have a well-grown shell but a minority have thin patches or rough interfaces, degradation will not be uniform. The vulnerable minority will generate traps sooner under illumination and oxygen exposure, will charge more readily under electrical stress, and will show faster non-radiative losses. Those failing particles then distort the ensemble spectrum and accelerate perceived dimming.

Ligand coverage adds a further layer of variability that is chemically subtle and operationally loud. Ligands are not arranged like tiles on a roof. They can pack densely on certain facets and loosely on others; they can be exchanged during processing; they can partially detach under heat; they can be displaced by components of a host matrix or by impurities. Two batches synthesized under nominally identical recipes can differ in how completely and how uniformly ligands passivate surfaces. That difference may not show up dramatically in an initial spectrum measured at low excitation in an inert environment. It can appear months later as a difference in shelf stability, or hours later under high photon flux or carrier injection.

Defect density is the axis that ties all the others together. A "defect" in this context is any structural or chemical irregularity that creates unwanted states where energy can be lost non-radiatively. Some defects are born during growth; others are created during processing; still others appear through aging. What matters for reliability is not only the average defect density but its spread. A few high-defect particles can dominate long-term behavior if they preferentially capture energy or charges. That is why reliability in devices is an ensemble property: lifetime scatter and yield depend on how consistently the emitter, ligand-related passivation, and interfaces are produced, not only on the best-case dot design.

There is a second, slightly uncomfortable reason sameness matters: it can masquerade as aging. A product can appear to "degrade" simply because its starting point was uneven. Imagine two display panels built from two batches of emitters. Batch A begins with a slightly broader size distribution and a slightly lower average quantum yield. Batch B begins narrower and brighter. Put both through the same operational test. Panel A may look as if it ages faster, not because its chemistry is fundamentally less stable, but because it started with a larger fraction of weak particles that were already near the edge. When those weak particles tip into non-radiative behavior, the early decline is steep. In a reliability report, that early decline can be misread as rapid wear-out rather than as variable starting quality revealing itself.

This distinction becomes commercially decisive. Companies do not merely need a long average lifetime; they need a narrow distribution of lifetimes across units. Consumers experience the tail, not the mean. A brand's reputation is shaped by the unlucky fraction of products that drift, dim, or fail early. Uniformity narrows that tail.

How, in practice, does sameness get lost? One culprit is scale. In small research batches, synthesis can be exquisitely controlled: fast heating, vigorous stirring, careful timing. Scaling batch synthesis tends to reduce control of fast nucleation and growth processes and thus uniformity and reproducibility; this is explicitly cited as a challenge in large-volume batch reactors. When a reaction happens in seconds, "mixing" is not an afterthought. In a small flask, reagents meet quickly and temperature equilibrates rapidly. In a large vessel, local conditions can differ: one region sees precursor earlier, another later; one region runs hotter; another is cooled by a wall. Nucleation may begin in one pocket while growth continues elsewhere. The result is a broader distribution in size and often a broader distribution in structural quality.

A historical anecdote from the culture of colloidal synthesis makes the point vivid. The hot-injection method, which became famous for producing narrow distributions in a laboratory setting, is built on a dramatic moment: precursors are injected into a hot coordinating solvent, and nucleation happens in a "burst". In a small flask, that "burst" can be close to simultaneous across the solution. In a larger reactor, injection is not a point; it is a process with spatial gradients. The very trick that produces monodispersity in the lab can become a source of heterogeneity at scale. Researchers scaling up often discover that the recipe is not a recipe; it is a choreography that depends on geometry.

That realization has pushed manufacturing toward approaches that treat mixing as a primary design variable rather than a background detail. Continuous or intensified mixing/reactor approaches (for example, impinging jet mixers) are pursued to improve reproducibility while reaching industrially relevant throughput. The idea is to create reaction environments where local conditions are kept consistent: reagents meet in a well-defined region with rapid mixing; residence times are controlled; heat is managed in a predictable way. Instead of scaling volume and hoping the same physics holds, the system scales by numbering-up or by maintaining the same mixing regime while increasing throughput. Sameness becomes an engineered output.

Chemistry itself offers knobs for reproducibility that are sometimes more powerful than mechanical control. Systematic synthesis control, such as modulating precursor reactivity, is positioned as a route away from trial-and-error and toward reproducible quality across sizes and materials. "Precursor reactivity" sounds like a specialist's phrase, but the intuition is straightforward: if the chemical building blocks are too eager, the reaction becomes a race that amplifies tiny differences in local conditions. If their reactivity is tuned so that growth is less explosive and more controlled, the system becomes more forgiving. Forgiving chemistry produces more uniform particles, and more uniform particles age more uniformly.

Uniformity also has a quieter benefit: it makes diagnosis possible. When a batch is highly consistent, changes under stress can be interpreted with more confidence. If a film dims uniformly, the mechanism is likely something that affects all particles similarly: perhaps a slow change in the host matrix, or a general increase in surface traps. If a film's spectrum broadens and its color point drifts, that suggests differential aging: some subpopulation is changing faster. Without a tight initial distribution, these interpretations blur. Heterogeneity is not only a reliability problem; it is a knowledge problem. It makes it harder to learn which protective strategies truly work.

Metrology, the practice of measuring what you have made, becomes part of the reliability story for the same reason. A narrow emission spectrum is not only desirable; it is also evidence that size distribution is tight. But many other forms of sameness are invisible to a simple spectrum. Shell thickness distribution, ligand coverage uniformity, and certain classes of defects may require complementary measurements. In manufacturing, quality control is often a game of finding fast proxies that correlate strongly with long-term performance. The ideal proxy is sensitive enough to catch the vulnerable minority before it leaves the factory.

A concrete product-facing example shows how this plays out. Consider a roll-to-roll process that produces a color-conversion film for displays. The emitters are blended into a polymer and coated onto a substrate. From a distance, the process looks like printing. Up close, it is an ordeal for a nanoscale surface: solvents, shear forces, drying gradients, possible exposure to oxygen, contact with additives in the polymer formulation. If the particles are uniform and robustly passivated, they will tend to remain dispersed and maintain their optical properties through processing. If a fraction of them has marginal ligand coverage, those may aggregate during drying. Aggregates can scatter light, broaden emission through reabsorption, and create local regions that heat under illumination. A customer will not say "your dispersion control failed"; they will say the whites look slightly off, or that the screen no longer matches another screen in the same room. The complaint is macroscopic, but the origin is a minority subpopulation that behaved differently during processing.

Sameness also interacts with the protection strategies described earlier. Encapsulation and barrier layers slow oxygen and moisture ingress, but if oxygen does get in, it will find the easiest targets first. A uniform population gives it fewer easy targets. A heterogeneous population hands it pre-made entry points: thin shells, defective interfaces, sparsely covered surfaces. In that sense, uniformity is a kind of passive protection, not by adding a new barrier but by removing weak spots.

There is a final irony. When a field is young, dramatic improvements often come from bold new structures: thicker shells, new ligands, new encapsulation strategies. As a field matures, progress is increasingly won by reducing variance. That kind of progress is harder to celebrate because it does not always come with a striking new image or a catchy name. It shows up as tighter bins on a quality chart, fewer outliers in a lifetime test, less color drift across a product line, and fewer returns.

A display that stays visually consistent for years is, in a quiet way, a triumph of sameness. It means countless microscopic interfaces behaved similarly, that few particles began life compromised, and that the inevitable assaults of heat, light, and oxygen met a population that responded not heroically but consistently. When reliability is the goal, consistency is a form of strength.

Chapter 7

The Screen Revolution You Barely Noticed

The upgrade that made living-room screens look startlingly richer was almost invisible: not a dramatic new display type, but a nanoscale color-conversion layer slipped into an existing stack. Behind that modest intervention lies a chain of ideas stretching from quantum-size-tuned glow to moisture barriers, manufacturing tolerances, and a new definition of what it means for a screen to show "real" color.

7.1 When Blue Light Learned New Colors

Most people think of a television upgrade as a sharp break: a heavier box replaced by a thinner one, or a new acronym printed on the bezel. The quantum-dot display boom arrived in a quieter way. The panel on the front of the TV stayed an LCD, with the same basic job it has always done: act as a fast, pixel-by-pixel set of shutters in front of a bright lamp. The trick was to change the lamp.

That lamp had already undergone its own transformation. Older LCDs relied on fluorescent tubes or clusters of white LEDs. "White LED," in many products, means a blue LED plus a yellowish phosphor that converts part of the blue into a broad, warm band. The mix looks white to the eye, but it is a sloppy kind of white: a wide smear of wavelengths, with plenty of light falling in spectral regions that the LCD's red, green, and blue color filters will later reject. It is like trying to paint with three carefully cut stencils while a floodlight sprays color everywhere; the stencils still work, but much of the light hits the blocked parts and turns into heat.

Quantum dots entered the display industry first not as tiny pixels, but as a way of improving that lamp without asking factories to rebuild the whole front panel. A sheet containing nanocrystals was placed in the backlight stack so that the backlight's "white" became more deliberately shaped. A strong blue component was kept, while two narrow additions were created: a red and a green, each centered where the LCD's color filters actually pass light efficiently. The LCD remained the same set of shutters, but it was now being fed a better palette.

The physical act behind this is photoluminescence: a material absorbs a photon, briefly stores that energy in an excited state, and then re-emits a photon as it relaxes back down. In a quantum-dot backlight, the incoming photons come from blue LEDs. Some of those blue photons simply pass through the film unchanged. Others are absorbed by dots designed to emit green; others by dots designed to emit red. The nanocrystals are not being asked to carry electrical current or endure the high fields of a pixel electrode. They are asked to do something they can do remarkably well: take light in, and give light back out, at a chosen color.

The engineering measure that decides whether this is a clever demo or a viable product is quantum yield. The phrase sounds like academic bookkeeping, but it is straightforward: out of all the photons absorbed, what fraction come back out as emitted photons? A quantum yield close to 1 (or 100%) means the dots behave almost like ideal converters. A lower yield means the film becomes a sponge for blue light: it absorbs photons and then wastes the energy as heat or non-radiative processes, dimming the backlight and raising thermal stress. For a consumer display, where brightness is purchased with power, and power becomes both electricity and heat, quantum yield is the difference between a premium feature and an expensive liability.

One reason the "blue-plus-dots" architecture spread quickly is that it fit the existing division of labor inside an LCD. The LCD panel manufacturers were already excellent at making large, defect-free arrays of transistors, liquid crystal layers, and color filters. Backlight suppliers were already excellent at making uniform sheets of light with diffusers, films, and reflectors. A quantum-dot film could be slotted into the backlight supply chain as a replacement for a sheet that was already there, rather than as a new pixel technology that would require retooling the panel itself. Evolution, not upheaval, is often the route by which a laboratory material reaches the living room.

The idea also matched a simple human desire: a screen that makes familiar content look more convincing without asking the viewer to learn new habits. Even before people learned phrases like "wide gamut," they knew the feeling of reds that looked like brick instead of neon, and greens that resembled leaves rather than highlighters. A better-aimed backlight is not a philosophical upgrade; it is a visceral one.

Once you picture the backlight as a light-shaping device, the appeal of nanocrystals becomes practical rather than mystical. Conventional phosphors can convert blue into other colors, but their emission tends to be broad: the "green" is a wide hump, and the "red" may leak toward orange. Quantum dots can be tuned to produce narrower peaks, and, crucially, those peaks can be placed where the LCD's red and green filters have their highest transmission. The color filters are still imperfect, and the LCD still wastes light because any shutter system must block what it does not want, but less of the backlight ends up in the wrong places to begin with.

That is why quantum dots, in most mass-market "QLED" televisions, do not form the pixel layer the way OLED materials do. To make a self-emissive quantum-dot display, you would need each pixel to be a controlled light source, driven electrically, with decades of lifetime, uniformity across millions of subpixels, and manufacturing yields high enough for large panels. Those goals are not impossible, but they are hard in ways that involve charge transport, device physics, and enormous process discipline. The backlight approach asks for something narrower and more achievable: photoluminescence from a film that can be manufactured separately, tested separately, and then laminated into a familiar optical stack.

It also isolates the dots from a hostile electrical environment. A quantum dot's glow depends on the careful handling of excited electrons and holes inside a nanometer-scale crystal and at its surface. When dots are driven electrically, the device must inject charges into them repeatedly and uniformly. When dots are driven optically, the blue LED does the energetic work at a distance. The dots absorb photons, not current. That separation helped industry adopt quantum dots early, while researchers continued pursuing fully emissive quantum-dot LEDs in laboratories and pilot lines.

A concrete moment when this invisible architecture became visible, briefly, was the marketing push around "Triluminos" televisions around 2013. The promise was not "a new kind of LCD," but a more vivid palette. The underlying idea, reported widely in consumer electronics coverage at the time, was a blue backlight combined with quantum dots acting as downconverters to generate cleaner red and green components. Different companies experimented with different physical placements—some put quantum-dot material close to the LED edge of a light guide, others moved toward film-based approaches that spread the optical load. What mattered to the viewer was simply that familiar content gained colors that felt harder to achieve with older backlights.

The adoption path also reveals something about how modern technology often advances: through parts that can be swapped without rewriting the entire product. A film is a particularly friendly industrial object. It can be made in large rolls, cut to size, laminated with barrier layers, and inserted into a backlight module with relatively minor changes. That is why quantum dots first succeeded as an ingredient rather than as a headline. The consumer might have bought "a brighter TV" or "a wider color TV," while the real change was a sheet of nanocrystals quietly doing spectral carpentry.

The details of photoluminescence also explain why this works so well with blue LEDs specifically. Blue photons carry more energy than green or red photons. When a green-emitting dot absorbs a blue photon, it can re-emit a green photon and dispose of the extra energy as heat within the material. Likewise, a red-emitting dot can accept blue and return red. The reverse is not possible: you cannot reliably take red photons and turn them into blue with the same passive process. So the backlight starts with the highest-energy visible LEDs that are cheap, efficient, and easy to manufacture at scale, then "steps down" to the other primary colors. In lighting, this stepping-down is common. In displays, quantum dots made the stepping-down more selective and more controllable.

Selectivity has a second benefit: it reduces the burden on the LCD's color filters. A color filter is not a perfect gate. Its "red" passband is a window with sloped sides: it transmits strongly in a region, transmits weakly near the edges, and blocks outside. If the backlight's red component is a wide hill, lots of that hill falls on the filter's shoulders and gets absorbed. If the backlight's red component is a narrow peak centered in the window, a larger fraction makes it through. That can be turned into either better color at the same power, or the same color at lower power, or some compromise between the two. Engineers do not get to choose only one virtue; they bargain among brightness, efficiency, heat, lifetime, and cost.

Quantum yield enters this bargain repeatedly. A high-yield dot emits most of what it absorbs, which helps both efficiency and thermal management. But yield is not an isolated property you look up in a table. It depends on how the dot is made, how its surface is treated, what it is embedded in, and what else is around it. The dot must be protected from oxygen and moisture, and it must be kept from clumping together in ways that cause scattering or reabsorption. The moment you turn a vial of glowing solution into a square meter of uniform film, you have created a new set of problems: keeping the dots dispersed, keeping the film clear, and keeping the optical properties stable as the device warms up and cools down over years.

There is also a subtler efficiency issue that appears as soon as you put converters into a closed optical box: light can be reabsorbed. A red-emitting dot might absorb some green, or absorb some blue that was meant to pass through. The film must be formulated so that the converters do their work once, cleanly, without turning the backlight into a maze where photons bounce, get absorbed, and lose their way. The display industry already knew this kind of system-level thinking from phosphors and diffusers, but quantum dots made the optimization more sensitive, because their spectral features are sharper and their surfaces more chemically active.

If this sounds delicate, it is. Yet the very delicacy is part of what made the technology feel like a small miracle when it worked. For decades, consumer displays improved through brute-force approaches: more backlight power, more aggressive image processing, thinner packages. Quantum-dot backlights offered a different kind of progress:

wasting less light before it ever reaches the filters. That is an almost moral form of engineering, the satisfaction of getting more out of what you already have.

The story also carries a lesson about language. "QLED" became a popular label, but in mainstream products it usually described an LCD with a quantum-dot-enhanced backlight, not a self-emissive panel. The distinction can feel pedantic until you realize how it shapes expectations. A self-emissive display can turn individual pixels fully off; a backlit LCD must always manage some leakage through shutters. Quantum dots, in the backlight role, do not change the fundamental black-level physics of LCD shutters. They change the spectral quality of the light that the shutters modulate. Many viewers still perceive the result as a major leap because color is emotionally powerful: it is tied to skin tones, sunsets, sports uniforms, animated films, and the subtle cues by which we decide whether an image feels expensive or cheap.

A final reason this development was "barely noticed" is that it asked for no new behavior. Nobody had to learn a new interface, or buy new content, or rearrange their living room. The quantum dots were sealed inside the stack, doing a job that looked, from the outside, like ordinary illumination. Yet the improvement they enabled—a backlight whose red and green were no longer broad guesses but carefully placed targets—quietly changed the look of screens across laptops, tablets, and televisions.

At the end of the chain sits a familiar moment: someone streaming a film, or editing photos, or watching a child's cartoon. The viewer's attention stays on the story. The engineering triumph is that the light arriving at their eyes has been sculpted at the nanometer scale, converted photon by photon, so that the old LCD shutters can perform with a palette that finally fits their filters.

7.2 Why Purity Beats Raw Power

Walk into a brightly lit electronics store and every screen seems to be competing in the simplest possible way: more light. Whites look whiter, sports look sunnier, and animated films appear to have been dipped in varnish. It is tempting to assume that brightness is the master control knob for image quality, and that everything else is a footnote.

Display engineers learn the opposite lesson early: brightness is only useful if it lands in the right places.

The reason is sitting in plain sight, disguised as something mundane. An LCD does not produce red, green, and blue light by itself; it starts with a backlight and then uses color filters to carve that light into three streams. Those filters are choosy. They do not accept "sort of red" or "a bit greenish." They pass a band of wavelengths and block the rest. When the backlight is spectrally sloppy, the filters behave like bouncers turning away a long line of would-be guests. The screen can be radiating a great deal of power, yet a sizable fraction never gets to the viewer because it arrives in wavelengths the filter rejects.

That is the first sense in which purity beats raw power: it turns more of the same electrical input into *usable* light, light that actually makes it through the panel's gates.

The second sense is more emotional. A bright screen with imprecise primaries can still look strangely flat. Reds become brick-and-orange blends; greens drift toward yellow; deep blues lose their snap. The image may look loud without looking real. What you feel as "vivid color" is not simply high luminance. It is luminance delivered at wavelengths that the eye interprets as distinct, saturated primaries.

A useful word here is *linewidth.* In spectroscopy, linewidth is a measure of how spread out an emission peak is in wavelength: narrow means the material emits light clustered tightly around a particular color; broad means the emission is smeared over a wider range. For displays, narrow red and green peaks are a kind of precision machining. They can be positioned so that the backlight's red energy falls squarely inside the red filter's passband, and likewise for green. When those peaks are broad, they spill into regions the filters will block or into neighboring channels that cause cross-talk. You pay for photons you never get to see.

If this sounds like an abstract, lab-bench distinction, it helps to picture a simple thought experiment. Imagine an LCD whose green filter is a window that admits only a particular region of wavelengths. Now consider two backlights that are equally bright in total. In the first, the "green" portion is a broad hill that extends well beyond the window's edges. In the second, it is a narrower peak tucked neatly inside the window. The second backlight can produce a greener-looking green at the same power, because more of its green photons survive the filtering step. Alternatively, at the same perceived green brightness, it can run the backlight at lower power. Either way, spectral precision buys you leverage.

This is why narrow spectral peaks matter because LCD color filters discard large portions of broadband backlight spectra; narrowing the red/green peaks increases usable light within the filter passbands and improves achievable gamut at comparable brightness/efficiency tradeoffs. It is also why so much of the display industry's excitement around nanocrystal converters was never about making the backlight merely "brighter." It was about making it *cleaner.*

Cleaner has a technical meaning here. A display's color gamut, as we saw, is the set of colors it can reproduce. If the red and green primaries are too broad, you cannot reach the most saturated corners of that space. Instead of landing on a vivid, well-defined red, you land on a slightly diluted version. Instead of a rich green, you get a green that feels as if it has been mixed with white. That dilution is not always noticeable in a casual glance, but it becomes unmistakable when you compare two screens side by side, or when a film cuts to a scene where color is the story: a red dress in warm light, a neon sign against a dark street, the emerald of a forest after rain.

Here, particle size control becomes an industrial superpower. A converter material made from nanocrystals is only as good as its uniformity. If a batch contains a wide distribution of sizes, it behaves like a crowd trying to sing in unison while each person chooses a different key. The combined emission widens. A more monodisperse batch (meaning the particles are tightly clustered in size) emits in a narrower band and hits the intended spectral target more reliably. That is not only about beauty; it is about manufacturing. A television brand wants one model to match the next, and a factory wants the first thousand units to match the last thousand. Narrow, repeatable peaks are a kind of quality control you can see with a spectrometer and feel with your eyes.

The push toward wide color gamut in consumer screens did not begin as a living-room luxury. It was tied to professional demands, and to an old frustration: cinema and photography were being created with color possibilities that home displays could not quite deliver. Digital cinema, for instance, adopted its own color targets (DCI-P3 is a widely cited one), and as studios learned to grade films for those targets, the limitation of the average TV became more obvious. Colors that were meant to be lush and distinct arrived softened. The viewer might not know the name of the standard, but they sensed the mismatch: the theatrical look did not survive the trip home.

That gap created a market opportunity. When premium televisions began advertising that they could cover "more of the cinema palette," it was partly marketing, but it also reflected a real technical constraint:

you cannot claim a wide gamut if your primaries are not spectrally well placed. Nanocrystal converters with narrow emission peaks became an efficient way to hit those targets without redesigning the front panel.

Standards can sound like paperwork, but they matter because they reward spectral accuracy. Rec.2020, often invoked in discussions of ultra-high-definition video, is a demanding target for primaries. It is not enough to be bright; the red, green, and blue must be well separated. In practice, many displays approach these targets imperfectly, and compromises show up in the emission spectra long before they show up in a glossy spec sheet. This is where the phrase "the spectra tell the story" earns its keep. You can put almost anything on a box; you cannot fake where your light actually sits.

At this point, a reasonable question arises: if narrow peaks are so helpful, why not make them as narrow as possible?

Because screens are not instruments; they are products constrained by human perception, optics, and power budgets. Narrowing an emission peak can introduce losses. Some of those losses are direct: the converter may re-emit fewer photons if its internal processes are not perfectly efficient, or if the surrounding materials absorb or scatter part of the light. Some losses are systemic: the LCD's own optics may not transmit extremely narrow peaks with perfect uniformity across viewing angles; polarizers and other films can have wavelength-dependent behavior. You can win a spectral battle and lose the full optical war.

There is also the eye itself. Human vision is not a simple light meter. It weights wavelengths unevenly, and its perception of brightness is entangled with color. A tiny shift in green can feel like a big change in brightness because our eyes are particularly sensitive in that region. That is why display tuning often feels like an art as well as a science: you are shaping photons for a biological sensor with quirks and nonlinearities, then asking a brain to interpret the result as "natural." Two screens with similar measured luminance can be perceived differently, depending on how the spectrum is distributed.

This leads to a more modern concept that marketers sometimes mention, but engineers live with: *color volume*. A two-dimensional gamut diagram answers a simple question: what colors can you make at a fixed brightness? Color volume adds a third dimension: what colors can you make across a range of brightness levels, including the bright highlights demanded by high dynamic range (HDR) content? A display might cover a wide gamut on paper but lose saturation in bright scenes because it runs out of headroom, or because its backlight and filters cannot maintain clean primaries at higher power. When a sunset flares, when chrome glints, when a stage spotlight blows out the whites, you learn whether the screen can keep colors from washing out.

Here the phrase "purity beats raw power" becomes sharper. If you try to achieve vivid color purely by brute force, you may succeed at moderate brightness and fail at high brightness. The brighter you drive the system, the more any inefficiency becomes heat, and the more the optical stack may begin to drift. Spectral precision helps maintain saturation as luminance rises, because less power is squandered in wavelengths that will be filtered away. In the language of HDR, good primaries help preserve "colorfulness" in highlights, not just in midtones.

A concrete example brings this down to earth. Consider a sports broadcast with intensely colored uniforms under harsh stadium lights. A screen with broad red and green primaries may make the field bright and the whites punchy, yet the team colors can look oddly similar from a distance, as if the palette has been compressed. On a screen with narrower primaries, those same uniforms separate more cleanly: the red jersey is unmistakably red, the green turf looks less yellowed, and the image retains its color identity even when the camera catches reflective helmets and bright signage. It is not that the second screen is simply brighter; it is that its brightness is better allocated across wavelengths.

The adoption of nanocrystal-enhanced backlights also exposed a tension between engineering truth and branding convenience. "QLED" became a popular umbrella term, but it does not guarantee a particular spectrum, a particular gamut, or a particular HDR performance. Two televisions can both carry the label and behave differently because the underlying backlight design, the converter formulation, and the color filters differ. Some models may prioritize maximum peak brightness for HDR impact; others may prioritize gamut coverage; others aim for cost. This is why modern debates tie back to measurement: "wide gamut" (2D coverage) is not the same as HDR "color volume" (3D), and marketing terms like "QLED" do not guarantee high performance across models.

This does not mean branding is meaningless; it often signals the presence of a real component and a real design intent. It does mean that the honest story of color lives in spectra: where the red and green peaks sit, how wide they are, and how they line up with the panel's filters.

There is also a quiet historical irony here. For many years, display progress was sold as a pixel-count story: HD, then 4K, then 8K. Yet many viewers who could not reliably distinguish 4K from 8K at typical viewing distances could immediately see the difference between a conventional backlight and a spectrally sharpened one. Color is a low-level feature of human experience, processed quickly and emotionally. It changes how faces look and how places feel. A slightly improved red can make a scene warmer; a cleaner green can make it fresher; a deeper blue can make night scenes more plausible rather than merely dark.

Engineers sometimes speak of "hitting the primaries," as if it were a target practice exercise. The phrase undersells the human side. Good primaries are not just points on a diagram; they are the foundation for everything the screen tries to persuade you of. The glow of a candle, the blush on a cheek, the dyed fabric of a costume, the warning light on a dashboard in a film close-up: each depends on the display being able to place energy at the right wavelengths and keep it there while the scene shifts from shadow to glare.

That is why the most consequential improvement in many modern LCDs was not a radical change in how the image is formed. It was a refinement in what the image is formed *from*: a backlight whose red and green are not broad suggestions, but carefully tuned notes, strong enough and narrow enough to make the filters' job easier and the viewer's experience richer.

7.3 The Most Important Place You Never See

A display that looks effortless from the sofa is a hostile workplace for delicate materials. Behind the glass sits a bright cavity where light is generated, shaped, reflected, filtered, and partially absorbed. Temperatures rise and fall with the content on screen and the season in the room. Adhesives cure. Plastics slowly relax. Minute traces of water and oxygen wander in from the edges. None of this is dramatic on any given day. Over years, it becomes the difference between a vivid screen that stays vivid and a premium feature that fades into a warranty claim.

That slow, unglamorous struggle is where nanocrystal color conversion either becomes an industrial technology or remains a laboratory trick. The backlight does not care how elegant a spectrum looked on a spectrometer last week. It asks a blunt question: will the converter emit the same red and green, at the same efficiency, after thousands of hours in a warm optical stack?

The unromantic hero of this story is packaging.

The converter used in many commercial LCD backlights is not a bare layer of nanocrystals exposed to air. QDEF is a multilayer laminate in which the QD-polymer matrix is sandwiched between barrier films; the barrier is essential because QDs are sensitive to moisture/oxygen, and the film must remain optically clear, flexible, and uniform. That sentence can sound like manufacturing trivia until you imagine what "sensitive" means in practice. Moisture and oxygen do not arrive as a visible splash. They seep. A microscopic pathway at the film's edge, a pinhole defect, a slight delamination after thermal cycling: any of these can become a slow leak, feeding chemical reactions that dull emission, shift color, or create scattering sites that haze the backlight.

Barrier performance is treated as a primary engineering constraint: transmission rates for water/oxygen and defect control (pinholes, edge seal, lamination quality) determine lifetime and color stability (e.g., white-point drift). This is why the best description of the barrier films is not "protective" in a casual sense. They are engineered as a kind of atmosphere control, a miniature vault built around an optically active layer that must remain pristine while being heated and illuminated for years.

A useful everyday comparison is food packaging for oxygen-sensitive snacks: a bag that keeps chips crisp by slowing the inward crawl of moisture and oxygen. The analogy is imperfect—a display demands far lower permeation and far higher optical clarity, and it must tolerate higher temperatures without clouding, yellowing, or warping. Still, the comparison captures the basic logic: the material's performance is as much about what you keep *out* as what you put *in*.

This is also where the placement of the converter layer becomes a strategic decision. Putting a converter near the LEDs can expose it to intense heat and high optical flux. Heat accelerates degradation pathways; intense excitation can stress surfaces and nearby polymers; local hot spots create non-uniform aging. A film-based approach spreads the optical load more gently across an area, and it can sit in a part of the backlight stack that is easier to cool and easier to isolate from the hottest components. Film-based integration can be "minimally disruptive" to BLU manufacturing because it can replace a diffuser sheet while keeping most other optical components unchanged. That phrase "minimally disruptive" is not just corporate soothing; it is a survival trait in consumer electronics. If the converter can be introduced without forcing an overhaul of the light guide, the reflector geometry, the assembly line, and the test equipment, it has a chance to scale.

Once inside the stack, however, the converter layer must behave like a good citizen in an optical ecosystem. Backlights are elaborate sandwiches of films: reflectors that push light forward, diffusers that homogenize it, prismatic sheets that steer it toward the viewer, polarizers that do the LCD's gatekeeping. A converter film cannot simply glow; it must do so without introducing new losses. If it scatters light too strongly, it can lower brightness and create non-uniform patches. If it absorbs wavelengths it should transmit, it can starve one channel and force compensating adjustments that reduce efficiency elsewhere. If it reabsorbs its own emitted light, it can broaden peaks or reduce output. The converter becomes part of an optical budget where every percent matters.

This is where "systems engineering" stops being a buzzword. The nanocrystal formulation, the polymer matrix, the barrier layers, and the rest of the backlight films all interact. A change that improves chemical stability might alter refractive index and increase reflections at an interface. A polymer that disperses nanocrystals nicely might have a tendency to yellow under blue light exposure. An adhesive that bonds well might outgas trace molecules that interact with the emissive layer. None of these effects is a catastrophe alone. Together, they determine whether the consumer sees the intended color after five years or a slightly pink-white that makes skin tones look tired.

Color drift is an especially unforgiving failure mode because it is hard to ignore. A backlight that simply dims might be masked by turning up brightness, until the power budget or heat limits intervene. A backlight whose spectrum shifts forces a different kind of compromise: whites acquire a tint, and the entire image feels "off." Many viewers cannot name the problem, but they sense it in faces and familiar objects. Engineers sometimes track this as "white-point drift," because the perceived white of the display shifts as the relative strengths or peak positions of the primaries change. When you are selling an experience that promises realism, a drifting white point is a betrayal that does not stay hidden.

A manufacturing vignette captures how tiny defects can become visible consequences. Consider an otherwise excellent film laminate that has an occasional microscopic defect in its barrier: a pinhole, or a region where the barrier thickness is slightly reduced. During final assembly, the film is cut, handled, laminated into the backlight, and sealed at the edges. In a perfectly made unit, the internal environment remains stable and the converter performs as designed. In the unit with the defect, moisture enters slowly, not enough to fog the whole film but enough to locally degrade emission. The first symptom might not be a dead screen. It might be a faint non-uniformity that only appears on a gray test pattern, or a subtle shift in white balance at one edge. In mass production, such marginal defects are the enemy: they do not fail immediately, they fail in the customer's living room.

It is easy to underestimate how much labor and ingenuity go into preventing that scenario because the consumer never sees the barrier layers. They are literally invisible when they do their job. Yet their specifications—how low the water and oxygen transmission rates must be, how robust the edge seals must remain through thermal cycling, how defect-free the lamination must be—can dominate the entire design. A spectacular nanocrystal recipe without an equally spectacular barrier package is like a sports car engine left outside in the rain. It may start on day one. It will not stay beautiful.

Heat, in particular, deserves attention because it is both unavoidable and uneven. The backlight is an energy conversion chain: electricity becomes blue photons in LEDs; some photons are converted into longer wavelengths; many photons are filtered or blocked by the LCD stack; the blocked portion becomes heat. HDR content pushes this chain harder, because bright highlights demand more backlight power. The warmest region may be near the LED strips or edge, while the center of the screen stays cooler. The converter layer must tolerate these gradients. It must also tolerate the daily thermal rhythm of a device that is turned on, used, turned off, and left to cool, thousands of times.

Thermal cycling is a quiet test of mechanical integrity. Materials expand and contract at different rates. A barrier film and a polymer matrix may not stretch identically. Over time, stress can accumulate at interfaces, encouraging microcracks or delamination that become pathways for ingress. The challenge is not only to design a barrier that is impermeable in principle, but to ensure that it remains impermeable after years of flexing, warming, and cooling.

Moisture and oxygen are not the only villains. Light itself can be.

Blue photons are energetic enough to trigger photochemical reactions in polymers and impurities. An optical stack that stays clear under room lighting can yellow slowly under constant high-flux blue illumination. Yellowing is not merely an aesthetic problem; it absorbs the very blue needed to pump the converter, reducing brightness and altering the balance among channels. This is why the choice of polymers, stabilizers, and additives is not a mundane materials detail; it is part of the color story.

Placement choices inside the backlight also reflect a philosophy of risk management. Some architectures concentrate converter material near the LED source, which reduces the area of converter needed but increases thermal and optical stress. A film-based architecture spreads converter material across a large area, which can lower local stress but demands exceptional uniformity across the entire sheet. Uniformity is harder than it sounds. Any thickness variation can cause slight color or brightness variation. Any aggregation of nanocrystals can create scattering, reducing clarity. The manufacturing goal becomes deceptively strict: a large, thin film must look like nothing at all, while doing everything.

The "minimally disruptive" aspect of the film approach has another hidden advantage: it allows converter films to be produced, tested, and qualified somewhat independently from the LCD panel. That separation supports industrial learning. Barrier film suppliers can improve permeation and defect control; converter formulators can improve stability and emission; backlight integrators can optimize optics and thermal management. Progress can happen in parallel. The display that ships is the product of these streams meeting in a stack, each constrained by the others.

A historical note helps explain why this systems thinking became urgent in the early 2010s. As consumer devices thinned and brightness increased, backlights became more compact and more stressed.

Tablets and thin laptops demanded high luminance from a small cavity, with limited space for heat spreading. That environment is unforgiving to any converter that is not exceptionally stable. Early commercial deployments of nanocrystal-enhanced backlights had to prove not only that the color looked impressive on day one, but that it stayed stable in exactly these cramped, warm designs. A spectacular store demo is easy; a stable product line is hard.

This is also why discussions of display materials sometimes sound more like discussions of packaging science than of photonics. People imagine breakthroughs as new emission colors or exotic nanostructures. Often the decisive breakthrough is a lamination process that eliminates pinholes, or an edge-seal method that stays tight, or a barrier stack whose permeation rate is low enough that the converter's chemical clock slows to a crawl. These advances rarely make headlines, but they change what is commercially possible.

There is a final, almost philosophical twist. The value of nanocrystal conversion lies in precision: narrow, well-placed peaks that make color filters waste less light and produce richer primaries. Yet that precision is fragile. A small chemical change at a surface, a slow oxidation process, a minor interaction with a polymer additive, can broaden an emission peak or shift it. In a laboratory, you can keep a sample under inert gas and measure it carefully. In a shipping product, you must build the laboratory conditions into the object itself. The barrier layers are not an accessory; they are the means by which the device carries its own controlled environment.

The best screens, then, are not just triumphs of pixels and algorithms. They are triumphs of invisibility: transparent layers that do not scatter, seals that do not leak, materials that do not yellow, and a converter film that continues to emit the same carefully aimed red and green long after the unboxing thrill is forgotten.

7.4 From Boutique Nanocrystals to Billions of Screens

In a research lab, a successful nanocrystal synthesis can be a small triumph measured in milliliters: a vial that glows brilliantly under a lamp, a spectrum with peaks in just the right places, a photograph that makes the sample look almost unreal. In consumer electronics, success is measured in something closer to freight: rolls of film that must be uniform across square meters, shipments that arrive on schedule, and performance that remains within tight tolerances after months in a warehouse and years in a living room.

The leap from boutique material to industrial platform is rarely glamorous, and it is almost never purely scientific. It is a change in *scale*, in *repeatability*, and in *accountability*. A company selling a premium television cannot say, "This batch came out a little redder than usual," any more than an airline can say, "This wing is slightly different from the last one." Screens are assembled from parts made in different places, by different suppliers, under strict cost pressure. To fit into that world, nanocrystals had to become something that could be made reliably, tested quickly, shipped safely, and laminated into devices without surprising the manufacturing line.

A useful way to frame the display revolution is that the core optical idea arrived early, but the market impact waited for the supply chain to catch up. Primary/industry sources chart a commercialization arc: early positioning of QDEF as scalable (2011), first high-profile commercial QDEF deployments in premium devices around 2013–2014, then rapid TV-scale adoption with notable cadmium-free milestones around 2015. Those dates matter not because they are mere history, but because they show the rhythm of technology transfer: a concept is declared "ready," then it proves itself in a smaller, premium product where costs are tolerated and volumes are manageable, and only then does it push into the brutal economics of big televisions.

That early positioning in 2011 was, in effect, a claim about manufacturability. It said: this is not only a clever way to sculpt light; it can be turned into a consistent component. By 2013–2014, premium devices such as high-end tablets and laptops provided a proving ground. These products demanded excellent color in a thin package, and they were sold to buyers who noticed and paid for that improvement. They also offered a practical advantage for an emerging materials technology: smaller screen area means less converter film per device, making supply constraints less punishing and enabling manufacturers to learn before scaling to living-room sizes.

Then came the television market, where the numbers become unforgiving. A modest change in a backlight's bill of materials, multiplied across millions of units, becomes an existential business question. A small yield loss in a film lamination step becomes a large financial problem. An instability that causes returns after a year becomes a reputational catastrophe. If nanocrystal conversion was going to be more than a premium curiosity, it needed to fit the industrial logic of consumer electronics: high yield, stable performance, predictable supply, and a cost that could compete with alternative backlight technologies.

The hard part is that "make more of it" is not a single challenge. Scaling required not only better dots (yield, stability, spectral tuning) but also supply chain readiness (film manufacturing, barrier production, quality control, and OEM integration into BLUs). Each of those phrases contains its own world of problems.

Start with the nanocrystals themselves. A backlight converter is not forgiving of variation. If the red peak shifts by a few nanometers, the color balance of the screen can change. If the linewidth broadens, the gamut advantage shrinks. If efficiency drops, the TV must run its LEDs harder to hit the same brightness, raising heat and further stressing the system. So the manufacturer does not merely need "red" and "green" nanocrystals. They need red and green with specified peak positions, specified linewidths, specified conversion efficiency, specified stability under blue excitation, and specified behavior in the polymer matrix that will become the film.

Achieving that at scale forces a new relationship between chemistry and measurement. In a lab, a chemist can adjust reaction time and temperature by feel, guided by experience and a few spectra. In production, the process must be controlled so that different operators on different days obtain the same result. That means standardized precursors, controlled mixing, controlled thermal profiles, and quality checks that are fast enough to apply to every batch. It also means accepting an industrial truth that scientists sometimes dislike: if a property cannot be measured reliably and quickly, it cannot be controlled in a manufacturing environment. Spectral peak positions, linewidth, and emission intensity become not just scientific observations but acceptance criteria.

The step from nanocrystal synthesis to converter film adds another layer of complexity. Making a film sounds like a straightforward act of mixing particles into a polymer and casting it. In practice, it is the point where good material can be ruined. The nanocrystals must remain well dispersed; clumps scatter light and reduce brightness. The film thickness must be uniform across a large area; thickness variations become color or brightness non-uniformity. The polymer must remain optically clear and stable under long-term blue illumination; any yellowing steals pump light and changes color balance. The barrier layers, as discussed, must remain intact; tiny defects can cause slow degradation that only shows up after extended use. Film manufacturing is where optical engineering, polymer chemistry, and defect control meet in a process that has to run continuously and cheaply.

At the scale of televisions, quality control becomes a discipline in its own right. A backlight film is not an artisanal object inspected by hand. It is a component that must pass automated tests, often under time pressure, with clear thresholds for pass and fail. The tests must catch problems that matter to the viewer: peak drift, reduced efficiency, haze, scattering, non-uniformity, edge defects. They must also catch problems that matter to the manufacturer: batch-to-batch inconsistency, shelf-life instability, handling damage. A supply chain that delivers excellent performance in prototypes but inconsistent performance in production is not a supply chain; it is a source of risk.

That risk is amplified by the way consumer electronics is assembled. A television is built from modules: LED strips, light guides, optical films, LCD panels, driver electronics. These modules come from specialized suppliers. The backlight unit (BLU) is integrated with the panel under production schedules that leave little room for special handling. A converter film must be compatible with the adhesives, temperatures, and mechanical stresses of assembly. It must survive being cut, laminated, and sealed. It must not shed particles, outgas troublesome molecules, or warp under heat. It must behave like every other piece of optically clear plastic in the stack, except that it is doing something far more sophisticated.

This is why the early market success of nanocrystal backlights mattered beyond the immediate consumer benefit. Once a technology is adopted by major manufacturers, it becomes legible to the industrial system. Suppliers invest in capacity. Process engineers develop specialized equipment. Standards and test methods mature. The technology stops being a novelty and starts being a platform, a tool that manufacturers can plan around.

The commercialization arc also shows how policy can act as a plot driver. Regulation and materials choices are intertwined: EU RoHS exemptions govern specific cadmium-containing QD uses and include explicit time horizons and alternative-availability arguments; this drove continued investment in heavy-metal-free alternatives such as InP, which often have broader linewidths and device-performance gaps relative to Cd-based QDs. These are not small details tucked away in legal documents. They shape the direction of industrial research.

Cadmium-based nanocrystals, particularly cadmium selenide systems, have long been prized for their optical performance: high efficiency, narrow emission, and tunability. They are also cadmium-containing materials, and cadmium is heavily regulated. For a technology aiming at billions of consumer devices, that creates a tension. Companies and regulators had to decide under what conditions such materials could be used, how they should be handled, and what timelines should apply. Exemptions, where they exist, tend to come with an implicit warning: do not assume this permission lasts forever. That warning, in turn, motivates investment in alternatives.

Indium phosphide (InP) nanocrystals emerged as a prominent cadmium-free candidate. They can produce red and green emission suitable for displays, but they often require extra engineering to approach cadmium-based spectral purity and efficiency. Broader linewidths mean slightly less perfect primaries; lower intrinsic efficiency or stability can demand more careful packaging and optical design to reach the same on-screen performance. This is a technical challenge, but it is also an economic one: any additional engineering increases cost. The market does not reward "cadmium-free" as an abstract virtue unless it can be delivered without sacrificing the qualities consumers notice most: brightness, color, and reliability.

A concrete milestone in this narrative is the public emphasis on cadmium-free nanocrystal televisions around 2015. That moment mattered because it signaled that heavy-metal-free alternatives were not only a research topic but a commercial reality. It also changed the competitive landscape. Once a major manufacturer markets cadmium-free performance, others must respond, and suppliers must offer compliant materials at scale. Regulation, marketing, and engineering begin to pull in the same direction.

This interplay helps explain a pattern that repeats in modern technology: the "best" material in the lab is not always the material that dominates the market. The market chooses a bundle of properties, some technical and some social. A slightly inferior spectrum may be acceptable if it avoids regulatory headaches and reduces supply risk. A slightly higher cost may be acceptable if it enables a premium product line. A material that is chemically elegant but difficult to scale may lose to one that is merely good but manufacturable.

Scale also changes how research is funded and what kinds of problems become prestigious. Once nanocrystal backlights became a real product category, the scientific frontier shifted. It became less about proving that nanocrystals can emit narrow red and green and more about making them robust: improving stability under continuous blue pumping, preventing photo-oxidation, suppressing subtle degradation pathways, reducing optical losses from reabsorption and scattering, and ensuring compatibility with polymers and barrier films. Reliability engineering, which can feel pedestrian compared to discovering a new effect, becomes central. In a living-room device, a 1% change per thousand hours is not a minor detail; it is the whole story.

The need for reproducibility also changed the relationship between companies and measurement. In the early days, a glowing vial could sell an idea to a partner. In mass production, partners want specifications, test protocols, and evidence. They want to know what happens after shipping, after storage, after operation in hot climates, after years of daily use. They want to know how the converter behaves when the television is driven hard by HDR content. That pressure forces a kind of maturity on the entire ecosystem, from synthesis to lamination to BLU integration.

There is, too, a human dimension to scaling that is easy to overlook. When a new materials technology transitions from lab to industry, it creates a demand for hybrid expertise. Chemists must learn the vocabulary of display optics and manufacturing yield. Optical engineers must learn enough chemistry to understand degradation pathways and packaging constraints. Quality engineers must learn to interpret spectra. Entire teams form around tasks that did not exist before: keeping emission peaks stable across batches, detecting barrier defects before lamination, modeling how a small change in converter output affects the final image after it passes through color filters and polarizers. The technology does not spread merely because it works; it spreads because people and organizations learn how to make it work *reliably*.

A useful anecdote from manufacturing culture captures the mindset shift. In a lab, a scientist can celebrate a batch that is "the brightest we've ever made." In a factory, the celebrated batch is the one that is indistinguishable from the previous hundred batches. That is not a rejection of innovation; it is the foundation that allows innovation to reach scale. Consumers do not buy a television hoping it is a rare, exceptional specimen. They buy it expecting it to match what they saw in the store and what their neighbor bought last week.

All of this helps explain why the screen revolution was so easy to miss. It did not arrive as a new display type in the way OLED did. It arrived as a new *ingredient* in an existing architecture, made viable by a chain of improvements that rarely appear on spec sheets: scalable synthesis, tight spectral control, stable film fabrication, robust barrier packaging, and production-friendly integration into backlight units.

The payoff is visible but understated. When a manufacturer can order converter film by the roll, with predictable peaks and stable performance, wide-gamut color stops being a boutique feature reserved for a few models. It becomes a design option across product lines. Better color moves from the realm of demonstrations to the realm of default expectations. The living room becomes a place where cinema color is routine, and where a thin sheet of carefully engineered nanocrystals quietly earns its keep, night after night, by doing the same precise job without complaint.

Chapter 8

Seeing Cancer in Technicolor

What if a biopsy could be read less like a grayscale map and more like a crowded night skyline, each color marking a different molecular clue? Here the reader discovers why quantum dots became such alluring biomedical labels—not just because they are bright, but because their size-tuned spectra, durable fluorescence, and engineerable surfaces make them unusually informative, even as the body stubbornly resists behaving like a laboratory vial.

8.1 A Dozen Signals in a Single Glance

A tumor is not a uniform lump of bad cells. Under the microscope it resembles a crowded city: cancer cells jostle with blood vessels, immune infiltrates, connective tissue, and pockets starved of oxygen. Pathologists have long known this. The frustration was practical. Many of the most useful molecular clues in cancer are not visible in ordinary stains; they must be *labeled* with fluorescent tags that attach to specific proteins. Yet for decades, fluorescence microscopy often forced an awkward kind of patience: add one label, image it, strip it or switch filters, add another, image again. The tissue stayed the same, but time, light exposure, and tiny shifts in alignment did not.

What made nanocrystal labels so tempting to biologists was not merely that they glowed, but that they offered a specific optical convenience: many distinct colors could be driven with the same lamp or laser, then separated cleanly at the detector. In fluorescence, the usual sequence is simple. You shine light of one wavelength to excite a molecule into an energetic state; it relaxes and emits light of a longer wavelength. The emitted light is what you measure. The trick is that the "where" and "how sharply" of that emitted light matters far more than it seems at first glance.

Traditional fluorescent dyes and many fluorescent proteins have emission spectra that look like wide hills rather than sharp spikes. Their light spills across a broad range of wavelengths. If you label a tissue with two dyes whose emission hills overlap, the detector is asked to do a delicate job: decide how much of the light in a given channel belongs to dye A and how much belongs to dye B. With enough overlap, that becomes guesswork dressed up as calibration. Biologists call this *bleed-through* or *crosstalk*. It is not only a nuisance; it can manufacture apparent biology. A cell can seem "double positive" for two markers simply because one bright label leaks into the other channel.

Nanocrystals changed the geometry of that problem. Their emission bands can be much narrower, closer to distinct peaks. Their absorption, meanwhile, is broad: a single excitation wavelength can energize several colors at once. Put those together and you get a laboratory convenience that feels like magic the first time you see it work. Instead of matching each color to its own specially chosen excitation source, you can illuminate the specimen with one beam and collect multiple, well-separated emissions. The microscope becomes less like an instrument that tiptoes from one fluorophore to the next and more like a device that listens for several distinct notes in the same chord.

This is not a claim that the nanocrystals produce perfectly pure colors, nor that biological samples cooperate. Tissue itself fluoresces faintly, especially in the blue and green; this *autofluorescence* acts like a low-level fog. Antibodies can bind a little where they should not. Filters are not infinitely sharp. Detectors respond differently across wavelengths. But narrow emission peaks buy you room to breathe. They reduce the fundamental spectral entanglement that makes multiplexed fluorescence so fragile.

The payoff shows up quickly in the kind of question you can ask. Consider what it means to "type" a cancer. A tumor might express a growth-factor receptor on its surface, activate a signaling pathway inside, and sit in a microenvironment rich in suppressive immune cells. None of those properties is adequately captured by a single marker. If you stain for only one protein at a time, you can count how many cells have it, but you lose the spatial logic: which cells have *both* markers; whether the receptor-high cells sit near blood vessels; whether immune cells cluster at the tumor margin or penetrate the core. Cancer is as much about relationships as it is about parts.

Multiplexing turns staining from a series of isolated snapshots into a measurement of a network laid out in space. Imagine a tissue section from a breast tumor. You might want to label epithelial tumor cells, dividing cells, blood vessels, cytotoxic T cells, macrophages, and a marker of low oxygen. Each label answers a straightforward question. Together they answer a different sort of question: where are the dividing tumor cells relative to oxygen-poor regions; do T cells reach those regions or stall at the edges; are macrophages positioned like gatekeepers at vessels where tumor cells could enter the bloodstream? Those are not abstract curiosities. They touch metastasis, treatment response, and the likelihood that an immune therapy will find its targets.

A well-told early story in this space comes from the period when nanocrystal bioconjugates were just becoming practical for biology. In 2004, Xiaohu Gao, Yuanyuan Cui, and colleagues demonstrated tumor targeting and imaging in living animals using nanocrystals coupled to biomolecular recognition elements. The headlines were about bright dots homing in on cancer. The deeper lesson was methodological: if you could build a panel of distinct colors that all responded to the same excitation and stayed spectrally separable, you could label several molecular addresses at once. In vivo proof-of-concept studies were necessarily simplified, but they crystallized an aspiration that pathologists immediately recognized: stop choosing which single marker to look at today, and start looking at a panel that reflects how tumors actually behave.

Multiplexing sounds like "more colors," yet the practical transformation is subtler. Ordinary immunofluorescence already uses multiple channels. The change is what happens when channels become numerous enough that you can treat them as data rather than decoration. At low multiplex (two or three markers), the mental model is qualitative: green cells here, red cells there, yellow where they overlap. At higher multiplex, the specimen becomes a map with many layers. You start to compute with it: classify cell types by combinations of markers; quantify distances between cell classes; measure gradients from vessels; cluster neighborhoods with particular immune compositions. The microscope image begins to resemble a geographic information system for tissue.

To make that work, the optics must not be the bottleneck. Narrow emission peaks reduce overlap, but do not abolish it. Real multiplexing relies on a marriage of hardware and analysis. One approach uses a set of carefully chosen filter bands, each collecting a slice of the spectrum where a particular label dominates. Another approach, often called *spectral imaging*, records a fuller spectrum at each pixel (or at least a richer set of wavelength bins) and then performs *spectral unmixing*: a computational separation of signals based on known reference spectra for each fluorophore. Unmixing is conceptually straightforward: if you know what each pure label looks like spectrally, and you measure a mixture, you solve for the combination that best matches the measurement. In practice it is sensitive to noise, background, and the uncomfortable fact that reference spectra can shift slightly in different environments.

That environmental sensitivity is not just a technical footnote. Biology is messy in exactly the ways that confound multiplexing. Antigens can be more accessible in one region of a tissue section than another. Fixation can alter epitope shape. Thick specimens scatter light, mixing photons from slightly different depths. A marker can be abundant but dimly labeled if the antibody struggles to penetrate. When multiplexing is advertised as "ten markers at once," the unspoken assumption is that all ten behave politely. They rarely do.

This is why experienced practitioners talk about multiplexing as an engineering discipline, not a shopping list of fluorophores. The first constraint is spectral: channels must be separable enough that unmixing is stable. The second constraint is photophysics: all labels must be bright enough to detect at reasonable exposure without saturating the brightest channels. The third is biology: the staining must be reproducible and specific, and the tissue must tolerate the combined protocol without losing morphology.

A useful way to see the spectral constraint is to think about two kinds of overlap. There is *emission overlap*, the classic bleed-through problem. Narrow emission helps here. But there is also *excitation overlap*: if your excitation light accidentally excites an unintended label, you create signal in the wrong channel. Broad absorption makes it easy to excite many labels at once, which is exactly the point, yet it means you cannot use excitation selectivity to rescue a poorly chosen emission set. Multiplexing with shared excitation works best when the emissions are well separated and the detection system can discriminate them cleanly.

Detector sensitivity brings its own bias. Many cameras and photomultiplier tubes respond less efficiently in the near-infrared. Tissue autofluorescence, on the other hand, tends to be weaker at longer wavelengths. So the near-infrared offers a quieter background but a detector that may be partially deaf. The "best" region of the spectrum is therefore a negotiated settlement among tissue, fluorophore, and instrument. This negotiation shapes what multiplex panels are practical in routine pathology, where the goal is not a heroic one-off measurement but a workflow that can be repeated day after day.

Multiplexing also forces a decision about what counts as a measurement. A single marker is often interpreted in a binary way: positive or negative. A multiplex panel invites more nuanced categories: cell types defined by combinations of markers; activation states defined by ratios; microenvironments defined by densities and distances. Ratios matter because they can cancel out some forms of variability. If two labels are delivered to a region of tissue with similar efficiency, the ratio of their signals can be more robust than either signal alone. This logic, familiar in other branches of measurement science, has begun to influence how multiplexed imaging panels are designed: not only to show where things are, but to support comparisons that are resistant to staining variability.

Yet there is a philosophical shift hidden inside the technical one. Under the old regime, the pathologist's gaze was selective. You chose a stain because you suspected something and you wanted confirmation. Under multiplexing, you can start with a broader, less committed measurement. The specimen can present surprises: unexpected immune cell neighborhoods; rare double-positive populations; spatial arrangements that suggest a mechanism rather than a label. That openness is not a guarantee of insight, but it changes the rhythm of discovery. It reduces the cost of asking, "What else is true in the same cells?"

A concrete illustration comes from the tumor microenvironment, where spatial relationships matter as much as molecular ones. Suppose a tumor is infiltrated by T cells, but therapy fails. A single-marker image might reassure you: "T cells are present." A multiplexed image can reveal the catch: the T cells sit in stromal regions and do not cross into tumor nests; macrophages form a band at the boundary; blood vessels are sparse in the core; hypoxia markers are high where T cells are absent. None of those observations alone is decisive, but together they outline a plausible story of exclusion, suppression, and poor access. This is the kind of story clinicians and researchers want: not merely what the tumor *is*, but how it is arranged, and therefore how it might be manipulated.

There is also a practical benefit that sounds mundane but becomes crucial in real laboratories: conserving tissue. Biopsies are small. A core needle sample may yield only a few thin sections. If each marker consumes a section, you run out quickly, especially once controls are included. Multiplexing lets you extract more information from the same slice of tissue, preserving material for other tests or for repeat analysis. That matters for patients as well as for science. Each additional biopsy is not just a data point; it is an invasive procedure with cost, anxiety, and risk.

The dream scenario is a panel that reads like a sentence: marker A tells you "tumor," marker B tells you "dividing," marker C tells you "immune attack," marker D tells you "immune suppression," marker E tells you "blood supply," marker F tells you "stress." Once you can read that sentence in one glance, you can start to look for grammar: patterns of co-occurrence and mutual exclusion; spatial punctuation at boundaries; rare "words" that predict response or relapse. The technical phrase *multiplexed measurement* sounds dry, but the emotional impulse behind it is humane. Cancer kills by being diverse within itself. Seeing that diversity, rather than averaging it away, is part of learning how to treat it.

Still, the ceiling is real. In a marketing brochure, it is easy to imply that if you can synthesize twenty distinct emission colors, you can measure twenty biomarkers at once. In practice, the limiting factors pile up: optical overlap, detector range, tissue autofluorescence, staining variability, and the sheer cognitive burden of validating a complex panel. A panel that is mathematically separable can still be biologically misleading if two markers interfere with each other's binding or if the tissue changes across the long protocol needed to apply many reagents.

That is why the most persuasive demonstrations of multiplexing are rarely the ones with the highest channel count. They are the ones where the additional channels answer a question that could not be asked convincingly before. A small increase in multiplexing power can unlock a qualitative change: the ability to identify a rare cell state by a combination of markers; the ability to locate that state within a specific niche; the ability to compare those niches across patients. When it works, it feels like switching from a black-and-white street photograph to a color-coded transit map: the same city, but now the routes are legible.

The optical trick at the heart of this approach is, in one sense, simple: shared excitation, narrow emissions, separable colors. The biological trick is harder: turning separable colors into separable meanings. When those meanings line up in the same tissue architecture, the tumor stops being a blur of "positive" and "negative" cells and becomes something closer to a system you can reason about—and, with luck, a system you can learn to disrupt.

8.2 The Light That Stays

A fluorescent label earns its keep in two currencies: how many photons it emits, and how long it keeps paying. The first currency is brightness. The second is endurance. In biomedical imaging, endurance often turns out to be the scarcer resource.

To feel why, it helps to picture a routine day in a pathology lab rather than a heroic microscope demo. A tissue section is stained with antibodies, each antibody carrying a fluorescent tag. The slide is then scanned, sometimes at high magnification across an area measured in square centimeters, not the tiny field of view of a single textbook micrograph. The scanner pauses, focuses, exposes, moves on, returns for another channel, perhaps revisits the same region to sharpen a boundary or resolve a small cluster of cells that looks suspicious. Photons are not just collected; they are spent. The act of observing is also the act of stressing the label.

Many conventional organic dyes respond to that stress with a slow collapse. They *photobleach*: their molecular structure is altered by illumination, typically through reactions involving oxygen or other chemical partners in the sample, and the altered molecule no longer fluoresces. In the resulting image series, the signal does not simply get noisier; it systematically shrinks. If you collect channel A first and channel B last, channel B may look fainter not because its target is rarer but because its dye has endured more light. That kind of bias is easy to overlook and hard to repair after the fact. Quantification becomes treacherous: the microscope has quietly edited the evidence.

This is where nanocrystal emitters gained their reputation for being unusually stubborn. A dye molecule is a small, specific chemical scaffold; when it breaks, it breaks. A nanocrystal is an extended piece of inorganic matter with many atoms sharing the burden of absorbing and releasing energy. That does not make it immortal, but it can make it much harder to destroy with light alone. Pathology-oriented workflows, in particular, reward labels that can tolerate extended illumination because they allow repeated scans, long exposures for dim targets, and careful re-imaging of the same tissue region without watching the signal dissolve.

Brightness and endurance intertwine but are not identical. Brightness is often summarized using *quantum yield* (as we saw earlier): the fraction of excitations that end up producing an emitted photon. A label can have a high quantum yield and still bleach rapidly. Another can have a modest quantum yield but endure so well that, over the course of a long scan, it yields more useful photons in total. For many imaging tasks, what matters is not "How bright is it on the first exposure?" but "How many photons can I collect before the specimen or the label gives up?"

The chemistry of photobleaching is a reminder that fluorescence is not just optics; it is also molecular wear and tear. Under illumination, a dye can enter reactive excited states that interact with oxygen to produce chemically aggressive species. Those species can then attack the dye itself. Biologists learn to fight back with mounting media, oxygen scavengers, and anti-fade additives, but these are palliative measures. They shift probabilities; they do not remove the underlying vulnerability of a small molecule repeatedly pushed into an energized state.

Nanocrystals can still lose brightness, especially in harsh conditions, but many are comparatively photostable because the emissive process is not tied to a fragile bond that must remain intact. Instead, much of the battle is fought at the surface: defects there can open pathways for non-radiative recombination, where excitation energy is released as heat rather than light. Shell growth and surface passivation were pivotal steps in making these particles behave as reliable fluorophores. The popular image of a "tiny lightbulb" is misleading; a better intuition is that a nanocrystal is a tiny stage where excited electrons and holes are trying to find each other and radiate, while surface traps are trying to steal them into silent exits.

Endurance matters in ways that are both mundane and profound. On the mundane side, whole-slide imaging is a grind: hundreds of fields, multiple channels, repeated autofocus routines. A bleaching dye forces a lab to reduce exposure times, increase illumination intensity (which can accelerate bleaching), or accept lower signal-to-noise. On the profound side, endurance changes what becomes thinkable. If a label can tolerate long observation, then you can attempt measurements that depend on time: tracking a moving cell, watching a tumor cell squeeze through a vessel wall, or following the flow of lymph toward a sentinel node. In those settings, a fluorophore that fades at the wrong moment does not merely spoil an image; it severs a narrative.

That narrative quality was part of the excitement around near-infrared nanocrystal probes in the mid-2000s. In 2004, near-infrared-emitting nanocrystals were demonstrated for sentinel lymph node mapping, an application that sits halfway between imaging and surgery. Sentinel nodes are the first lymph nodes to which a tumor is likely to spread. Surgeons want to locate them quickly and remove them selectively. Traditional dyes can help, but near-infrared light penetrates tissue more effectively and is less confounded by the tissue's own autofluorescence. A bright, stable near-infrared label that remains emissive in biological fluids offers a kind of real-time guidance: lymphatic channels can be followed, nodes can be identified against a darker background, and decisions can be made with more confidence about what to sample. The appeal here is visceral. You are not making a picture for a paper; you are trying to see anatomy and disease in the living body while time matters.

There is a catch, and it is worth taking seriously because it reveals a deep truth about measuring with light: the behavior of a fluorophore in a bulk sample can hide the unruly behavior of single emitters. Under a microscope, a tissue region may contain thousands or millions of labeled molecules or particles. Their emissions average together. If some fraction flickers off or fails temporarily, the ensemble signal can remain smooth. The image looks stable, and it is, in aggregate. But when the task depends on individual particles or small clusters—for instance, tracking a single labeled receptor, following an individual cell, or performing super-resolution microscopy where each blinking event can be used as a localization cue—the story changes.

The field has a name for the stubborn intermittency of single nanocrystal emitters: blinking (defined earlier). The practical meaning is simple and maddening: a label can remain chemically intact and photostable over long durations, yet its emission can switch between bright and dark states on timescales from milliseconds to seconds. From the outside, it resembles a faulty bulb. From the inside, it is a competition between radiative pathways (light-emitting) and non-radiative pathways (energy-dissipating), with the balance tipped by subtle changes at the surface or in charge state.

This is the moment where "the light that stays" becomes a more complicated promise. Photostability can mean that the emitter does not permanently burn out. Blinking means that it may still abandon you intermittently. For many ensemble imaging tasks—staining a tissue section, mapping marker expression, building multiplexed panels—intermittency is softened by numbers. Enough labels are present that the camera rarely sees a complete blackout. For single-particle tracking, intermittency can break trajectories into fragments and bias the measurement toward the longest "on" times. The same physical object can be both a triumph of endurance and a nuisance of caprice.

Why does blinking happen? In broad brushstrokes, it is connected to how charges move and get trapped at the nanocrystal surface or in its shell. If an extra charge is present, a newly created electron—hole pair can recombine without emitting light, transferring its energy to the extra charge instead. This non-radiative process is often associated with Auger-related loss pathways, named after Pierre Auger, and it is ferociously efficient at turning what would have been a photon into heat. The nanocrystal is not "damaged" in the way a bleached dye is damaged; it is simply in a state where light emission is temporarily outcompeted.

The unsettling implication is that there is no single, universal photostability rating for a nanocrystal probe. Single-particle investigations have reinforced that stability is conditional: buffer composition, oxygen levels, illumination intensity, and the presence of reducing agents can shift the balance between emissive and dark states. A particle that behaves impeccably on a glass slide in one mounting medium may flicker more in a different chemical environment, or under a different excitation regime. "Photostable" is therefore not a badge; it is a performance that must be verified in the specific context where it will be used.

This conditionality collides with a common expectation in biomedicine: that a label is a passive reporter. The mental model is a sticky note attached to a molecule, saying "Here I am." In reality, fluorescent labels are participants in the measurement. They absorb energy, enter excited states, exchange charges with their environment, and sometimes catalyze small chemical changes around themselves. With dyes, that participation often ends in photobleaching. With nanocrystals, it can manifest as intermittency, spectral shifts, or subtle intensity changes over time.

A second complication is that the kind of illumination used in modern microscopy can be harsh in ways that old fluorescence images rarely faced. Confocal microscopes concentrate light into small volumes and can drive high excitation intensities. Super-resolution techniques may deliberately push fluorophores through cycles of dark and bright states to achieve nanoscale localization. High-speed cameras encourage high frame rates, which can tempt the experimenter to increase illumination to keep each frame bright. In that setting, photostability is not a luxury; it is a precondition for obtaining enough photons to do the math.

Here is the paradox that many biologists encountered when they first adopted nanocrystal labels: the slide could be scanned again and again with little sign of fading, yet at the single-particle level the signal could behave like a strobe light. Both observations were true. They simply lived at different levels of description. It is the same gap that appears in many sciences: a material can have stable bulk properties while its microscopic constituents remain restless.

Workarounds emerged alongside the frustration. Improved shell designs and surface treatments reduced blinking for many formulations. Experimentalists learned to tune imaging conditions to suppress long dark states, at least partially. For applications that depend on ensembles, the community learned to embrace the steadiness of the average and treat single-particle quirks as secondary. For applications that depend on single emitters, others did the opposite: they turned blinking into a feature, using intermittency as a way to isolate individual emitters temporally so they could be localized more precisely. A behavior that looked like unreliability could, in the right framework, become a tool.

Underneath these tactical choices sits a broader reason endurance mattered so much in cancer imaging. A tumor's most informative features are often subtle: a thin rim of invasive cells; a sparse population of immune cells that predict response to therapy; a small cluster of cells with a stem-like signature that seeds relapse. Detecting subtle features requires either very high contrast or very long observation, often both. If the label fades as you search, you are fighting a clock. If the label holds, you can keep looking until you are satisfied that a rare signal is real and not a noise spike.

There is also a psychological dimension that rarely makes it into methods sections. When you image a precious clinical sample, you want to be able to return to it. A pathologist may notice a pattern only after seeing another region. A researcher may reanalyze an image with a new algorithm months later and wish they had collected a slightly different set of exposures. A stable fluorophore does not solve every problem, but it reduces regret. It gives you permission to take the time the specimen deserves.

The enduring glow of nanocrystal probes, then, is not just an aesthetic property. It is an argument about what should count as a measurement in biology. A measurement is not a single photograph; it is a process that unfolds under repeated illumination, across fields of view, across channels, across time. Endurance keeps that process honest. It makes it harder for the instrument to quietly rewrite the sample by dimming it.

And yet, the same physical mechanisms that grant endurance can introduce their own kind of uncertainty. Photobleaching is a failure you can see and name: the signal decays, you stop, you adjust. Intermittency is more insidious. A label can be present but dark, a silent witness. If your experiment assumes that "no light" means "no target," blinking can masquerade as biology. The remedy is not cynicism; it is calibration, replication, and an attitude toward fluorescent signals that treats them as the outputs of complicated objects rather than perfect little lamps.

A cancer cell does not care how beautiful your image is. It cares whether it can divide, migrate, evade immune attack, and find a blood supply. To watch those behaviors, the scientist needs light that does not tire easily, because the questions are not answered in a single flash. The most valuable glow is the one that persists long enough for the biology to reveal its timing, its patterns, and its exceptions.

8.3 Teaching a Nanocrystal Where to Go

A glowing nanocrystal on a benchtop is an optical object. A glowing nanocrystal in a tissue sample is a social object: it meets proteins, membranes, sugars, salts, and the crowded, sticky complexity of biology. Whether it becomes a helpful beacon or a nuisance depends far more on what is attached to its surface than on the semiconductor at its center. As we saw with "packaging as hero," the business end of many nanomaterials is not the core that physicists celebrate, but the interface that cells actually touch.

That interface has two jobs that pull against each other. First, it must keep the nanocrystal bright by preventing the surface from becoming a sink for non-radiative loss. Second, it must make the particle behave in water, in salt, in serum, in the presence of cells, and in the presence of whatever biological targeting molecule you hope will guide it. A nanocrystal that glows beautifully in an organic solvent can become useless the moment it meets a physiological buffer: it can clump, stick nonspecifically to everything, or simply stop emitting as the surface chemistry changes.

The phrase "teaching a nanocrystal where to go" is a convenient anthropomorphism, but it captures the practical problem. The body does not want your particle. A microscope slide does not care what your particle does. In between sits the craft of surface engineering: layering chemistries so that an inorganic crystal can remain colloidally stable (well-dispersed) and optically competent while carrying a molecular "address label" that recognizes a specific protein or cell type.

A useful way to think about a biomedical nanocrystal probe is as a layered object. The core determines the color. A shell often improves brightness and stability by smoothing away surface imperfections that would otherwise swallow excitations without emitting light. Then comes the part biologists obsess over: the coating and its outermost functional groups. That outer layer decides whether the particle stays dispersed in salty water, whether it binds to proteins nonspecifically, how big it appears in solution, and whether it can carry a targeting molecule without losing its manners.

The first challenge is simply getting the nanocrystal into water. Many high-quality nanocrystals are made in hydrophobic solvents with hydrophobic ligands on their surface. Those ligands are excellent for synthesis: they bind to the growing crystal, control its size, and keep it suspended in organic liquids. They are also a disaster for biology, which is aqueous. A hydrophobic particle dropped into water tends to aggregate, like oil droplets merging. Aggregation is not a minor inconvenience. A clump behaves as a different object: it scatters light, settles out of solution, becomes easier for cells to engulf, and may lose the very spectral properties that made it attractive.

Turning a hydrophobic nanocrystal into an aqueous probe is often done in two broad ways. One is *ligand exchange*: replacing the original hydrophobic ligands with molecules that still bind the surface but present hydrophilic groups to water. The other is *encapsulation*: wrapping the particle, original ligands and all, in an amphiphilic coating whose inner face likes the hydrophobic surface and whose outer face likes water.

Each route carries trade-offs that become painfully real in the lab. Ligand exchange can reduce the distance between the emissive core and the outside world, keeping the overall size small. That can be valuable for tissue penetration and for any later attempt at clearance from the body. But exchanging ligands can also expose or create surface defects that reduce brightness. Encapsulation can preserve optical performance by leaving the original surface chemistry largely intact, but it usually adds bulk. In biology, bulk is not an abstract number; it changes how the particle diffuses through tissue and how it is taken up by cells.

Once the particle is water-compatible, the next challenge is to keep it from behaving like lint. Biological fluids are full of proteins evolved to stick to things. A bare or poorly coated nanoparticle is a tempting landing pad. Proteins can adsorb rapidly, building a layer that changes the particle's effective size and its surface character. This protein layer can be helpful if you want the particle to be eaten by immune cells, but it is usually the opposite of what you want for targeted imaging. It also creates maddening batch-to-batch variability: two samples of "the same" nanoparticle can behave differently because of small differences in surface chemistry that steer protein adsorption.

Steric stabilization is one response. By attaching polymer chains to the surface, you create a soft brush that makes it energetically unfavorable for particles to come close enough to stick together. Poly(ethylene glycol), often shortened to PEG, became a widely used choice because it is hydrophilic and tends to reduce nonspecific binding. In the language of nanomedicine, it can provide "stealth," discouraging proteins and cells from treating the particle as an invader. Stealth is never perfect, but coatings can shift the balance: from immediate sequestration and sticky artifacts toward longer circulation times and cleaner staining.

Now comes the step that gives the probe its purpose: bioconjugation, the attachment of a targeting molecule. Here, the biologist's favorite tools enter: antibodies, peptides, oligonucleotides, aptamers, and small molecules. The guiding idea is simple. If you can couple a nanocrystal to something that naturally recognizes a biomarker, the fluorescence becomes a readout for the biomarker's location.

In practice, "couple" hides a world of chemistry. A targeting molecule is not merely a hook. It is itself a large, charged, flexible object that changes what the particle looks like to the environment. Adding an antibody can dramatically increase hydrodynamic size (the size the particle effectively has in solution), which can reduce tissue penetration and alter biodistribution. Adding many antibodies can increase binding avidity, but it can also encourage aggregation, increase nonspecific interactions, and make the particle more visible to the immune system. These are not abstract design constraints; they often decide whether a staining protocol yields crisp cellular outlines or a frustrating haze of background.

The most common targeting molecules in laboratory immunostaining are antibodies, for good reason. They are highly specific, widely available, and familiar to every pathology workflow. An antibody against a tumor-associated antigen, conjugated to a fluorescent label, can highlight malignant cells against a complex background. Early in vivo targeting demonstrations in the mid-2000s treated the nanocrystal as a platform: core and shell for optics, coating for biocompatibility, antibody for recognition. The conceptual shift was powerful. If the optics were good enough, you could build multifunctional probes: one particle, one excitation source, multiple colors, multiple targets.

But antibodies are not always the best teachers for a nanocrystal, because they are large. A full-size antibody is a bulky Y-shaped protein. When tethered to a particle that already has a coating, it can create a probe that is surprisingly big. In tissue, size slows diffusion and limits access to tightly packed regions. In living organisms, larger hydrodynamic size can steer probes toward uptake by the liver and spleen. These constraints motivate alternatives.

Peptides are a popular alternative: short chains of amino acids that can be designed or selected to bind specific receptors. They are smaller than antibodies, often more robust under varying conditions, and easier to attach in controlled numbers. Small-molecule ligands can be smaller still, sometimes binding to receptors or enzymes with high specificity. Aptamers, nucleic-acid sequences selected to bind targets, occupy another niche: they can combine the programmability of nucleic acids with binding behavior reminiscent of antibodies, while still being relatively compact.

Choosing among these options is a design problem with no universal answer. The "best" targeting strategy depends on what you need to measure. If the goal is to stain a fixed tissue section, penetration depth may be modest and size may be tolerable; specificity and low background may dominate. If the goal is to target a probe in vivo, the same size can become a liability, and the chemistry must anticipate circulation, protein adsorption, and clearance.

Here it is useful to separate two ideas that are often blurred together in popular descriptions: *active targeting* and *passive targeting*. Active targeting means specific binding: the probe carries a ligand that recognizes a biomarker. Passive targeting refers to the tendency of certain particles to accumulate in tumors because tumor vasculature can be leaky and lymphatic drainage imperfect. Passive accumulation can increase local concentration without any molecular recognition. Active ligands can then improve retention or uptake by particular cells within that tumor region. The two mechanisms can cooperate, but they can also be mistaken for each other. A probe may appear "targeted" simply because it accumulates nonspecifically in tumors and then sticks.

That ambiguity is another reason surface chemistry matters. A truly targeted probe should not merely accumulate; it should do so with a pattern that matches the biology of the biomarker, and it should do so reproducibly. Achieving that depends on suppressing the many nonspecific ways a nanoparticle can adhere to tissue: electrostatic attraction to charged surfaces, hydrophobic interactions, protein adsorption that creates new sticky patches, and aggregation that turns a probe into debris.

Ligand density adds another layer of subtlety. Intuition suggests that more targeting ligands should produce stronger binding. Often it does, up to a point. But increasing ligand density can change the probe's charge and increase its hydrodynamic size. It can also crowd ligands so that they interfere with each other's ability to reach their receptor. A crowded forest of antibodies can become a brush that prevents close approach. There is also the issue of orientation. An antibody only binds well if its binding sites are accessible; attach it randomly and you can end up with many antibodies that are technically present but functionally mute.

This is where the craft begins to resemble cooking more than engineering. Protocols are full of choices that look minor on paper: which linker chemistry, what buffer, what pH, how long to react, how thoroughly to purify. Those choices can change colloidal stability, brightness, and nonspecific binding. Bioconjugation changes more than "binding." It reshapes the particle's identity.

There is a final tension worth making explicit: the same surface that must interact specifically with a target must also resist interacting with everything else. Specificity is not simply a matter of a good antibody. It is a signal-to-background problem, and background is largely written by surface chemistry. In a tissue section, background can mean a diffuse glow in extracellular matrix, or bright speckles where particles aggregate, or uptake by cells that are simply good at engulfing foreign objects. In vivo, background can mean persistent signal in the liver and spleen that overwhelms the region you care about.

When the engineering is done well, the result can feel almost unfairly powerful. A tumor antigen that is invisible in ordinary stains becomes a crisp pattern. A rare population of cells becomes countable. Multiple markers can be read in the same specimen, and their spatial relationships become measurable. Yet the most honest view is that the nanocrystal is not a self-sufficient beacon. It is a component inside a larger device built from chemistry, biology, and optics, and the outermost few nanometers often decide whether the whole contraption succeeds.

The lesson is oddly human. Teaching is not accomplished by shouting louder; it is accomplished by making yourself understandable in the listener's language. A nanocrystal speaks photons. A cell speaks surface chemistry. The art is to build an interface where both languages can be translated without distortion, so that when the probe finally binds to a receptor on a cancer cell, the glow feels like evidence rather than accident.

8.4 The Body Is Not a Beaker

A stained tissue slide is a controlled world. The probes arrive by diffusion through a thin section, the wash steps are timed, and the microscope politely reports whatever light survives the protocol. Inject the same kind of glowing probe into a living body and control becomes negotiation. Blood flows, organs filter, immune cells patrol, proteins crowd every surface. The very properties that make an engineered nanocrystal such a satisfying object in the hand—a defined size, a tunable color, a deliberately designed coating—suddenly become variables inside a much larger system.

Pharmacologists have a tidy name for the life story of any injected agent: ADME, for absorption, distribution, metabolism, and excretion. It sounds like paperwork. In practice it is the difference between a probe that illuminates a tumor cleanly and one that vanishes into the liver, brightening the wrong anatomy with great enthusiasm.

Start with a simple thought experiment. Suppose a clinician wants to see whether a tumor expresses a particular receptor, and a researcher has built a nanocrystal probe coated for stability and decorated with a targeting ligand. In a beaker, the probe meets the receptor and binds. In the bloodstream, the probe meets everything *before* it meets the receptor: albumin and antibodies, lipoproteins and clotting factors, vessel walls and wandering macrophages. Each encounter offers an opportunity for the probe to be altered, captured, or redirected.

A central actor in this story is the *protein corona*. The moment a nanoscale surface enters serum, proteins adsorb onto it, forming a layer that can be thin and dynamic or thick and persistent. The corona changes what cells "see." It can mask the intended targeting ligand, reduce binding to the desired receptor, and add new unintended interactions. It can also make the particle effectively larger. In a test tube, you can report a particle's size precisely. In blood, the hydrodynamic diameter can swell as proteins cling, and that swollen size controls which biological filters the particle can pass.

Size governs fate in the body with a bluntness that can feel unfair. A key hard number emerges from renal physiology: efficient urinary excretion of inorganic nanoparticles in rodent models requires keeping the *hydrodynamic diameter* below roughly ~ 5.5 nm. Above that threshold, the kidney's filtration barrier tends to refuse passage; particles remain in circulation longer, increasing the chance of capture by other organs. The number is not a moral judgment; it is architecture. The glomerular filter is built to let small molecules through while retaining larger proteins, and it treats many nanoparticles as oversized guests.

This is where surface chemistry and physiology collide. A nanocrystal core might be only a few nanometers across, temptingly close to the renal-clearance regime. But the core is never injected naked. It wears a shell, then a coating to make it water-compatible, then often a polymer brush for stability, and finally a targeting ligand. Each layer adds to hydrodynamic diameter. Worse, a particle designed to be just under the cutoff in water can drift over it in serum because the protein corona inflates the effective size. The kidney does not care what you intended; it filters what you delivered.

Coating chemistry can shift that outcome. Zwitterionic or neutral coatings tend to reduce protein adsorption and can help preserve clearance behavior, precisely because they offer fewer attractive sites for proteins to land. A sticky or charged surface, by contrast, invites a corona and invites capture. Here "charge" is not a single value but a surface landscape: patches of positive and negative, local hydrophobic regions, and flexible ligands that can rearrange. Biology samples that landscape with relentless sensitivity.

When a particle does not clear through the kidneys, it usually meets the body's other major sorting machinery: the mononuclear phagocyte system (often described in older language as the reticuloendothelial system). Macrophages in the liver and spleen specialize in removing foreign objects from blood. They are not malicious; they are conscientious. Their job is to keep the bloodstream from becoming a distribution channel for pathogens and debris. Nanoparticles, no matter how elegant, can look suspiciously like both.

The consequence is a common pattern in biodistribution studies: accumulation in liver and spleen. For imaging, that can mean high background signal and a reduced amount of probe reaching the intended target. For safety, it raises a longer-term question: how long does the material persist, and what happens as it ages in tissue? A glowing signal is a convenience for the experimenter, but it is also a reminder that the material is still there.

All of this turns "targeting" into a more constrained concept than it appears in celebratory demonstrations. A targeting ligand can increase binding to tumor cells that a probe actually reaches, but reaching the tumor is itself a delivery problem. Tumors are irregularly vascularized. Their interstitial pressure can hinder entry. Their extracellular matrix can be dense. Some probes accumulate passively in tumors because tumor blood vessels can be leakier than normal ones, but that passive accumulation is variable across tumor types and across patients. Active targeting can then help with retention or internalization, yet it cannot rescue a probe that is cleared too quickly, sequestered too early, or blocked at the tumor boundary.

The simplest way to see the tension is to look back at the design choices from the bench. Increasing ligand density can improve the probability of binding once the probe encounters its receptor. At the same time, more ligands and thicker coatings increase hydrodynamic size and can shift biodistribution toward liver and spleen, while pushing the particle farther from the kidney-clearable regime. The body forces engineers into a three-way trade-off among specificity, circulation, and elimination. One can optimize any two; the third complains.

A concrete example of this tension appeared early in the history of in vivo nanocrystal imaging. In the mid-2000s, nanocrystal probes were demonstrated in animals for cancer targeting and for near-infrared imaging tasks such as sentinel lymph node mapping. The appeal was obvious: bright emission, reduced tissue autofluorescence in the near-infrared, and enough photostability to follow dynamic processes. But the same studies also highlighted the translational gap between an animal proof-of-concept and a human clinical agent. In mice, a probe that accumulates in liver and spleen may be acceptable for a short-term experiment. In patients, persistence and heavy-metal content invite a different standard of caution.

Cadmium-based nanocrystals sit at the center of that caution. Cadmium is a toxic metal, and cadmium-containing cores have therefore carried persistent safety concerns. Toxicity is not a single property; it depends on dose, duration, degradation, and where the material goes. A well-passivated particle may keep its cadmium sequestered for a long time, while a degraded or poorly protected one may release ions that participate in harmful chemistry. The toxicity literature emphasizes organ-specific outcomes and oxidative stress mechanisms, but it also carries an uncomfortable message: comparability across studies is limited. Change the coating, the size distribution, the animal model, the route of administration, or the illumination regime, and the outcomes can shift. The heterogeneity of materials and methods makes sweeping statements risky.

This is one reason the field has been motivated to explore cadmium-free alternatives, including indium-based nanocrystals, particularly for near-infrared emission. Substituting a different element is not a magic eraser. It can reduce a specific hazard while introducing new questions about biotransformation, persistence, and clearance. What matters to the patient is not whether the core contains cadmium or indium in the abstract, but how the complete particle behaves after injection: where it accumulates, how long it stays, what it becomes as proteins and cells process it.

Beyond toxicity lies a quieter barrier: translation is not only biology; it is manufacturing and regulation. A clinical imaging agent must be reproducible at scale, stable on the shelf, consistent from batch to batch, and characterized in ways that satisfy regulators who are trained to distrust cleverness. Nanoparticles complicate every one of those requirements. Their properties depend on distributions (not one size but a spread), on surfaces that can evolve during storage, and on small variations in synthesis that can have large biological consequences. A probe that performs brilliantly in a research lab can stumble on the unglamorous steps between the lab bench and a vial that a hospital pharmacy is willing to stock.

These realities do not erase the scientific value of in vivo nanocrystal imaging; they define it. The most durable contribution has often been in illuminating biological processes in animal models and in ex vivo human samples, where the goal is insight rather than approval. Tracking how tumor cells move, how immune cells arrange themselves at a boundary, or how lymph drains from a tumor region can shape the way researchers think about metastasis and treatment response. In those contexts, a bright, stable near-infrared label can reveal anatomy and dynamics that are otherwise difficult to see.

Still, the body keeps its veto power. Injected probes must contend with filtration thresholds, protein coronas that rewrite surface identity, and immune surveillance that mistakes "engineered" for "invading." The glowing core can be exquisitely engineered and still end up illuminating the liver simply because the liver is doing its job. That is the sobering lesson that follows any attempt to turn a nanocrystal into a medical tool: the same surface modifications that help a probe find disease also shape where else it goes, and how long it stays there.

Chapter 9

Harvesting Sunlight with a Tunable Palette

At first glance, sunlight seems too broad and unruly to be tailored. Yet quantum dots turn color into a design tool, inviting a new way of thinking about energy capture in which nanocrystals are tuned not just to shine beautifully, but to harvest, redirect, and survive the light that falls on them. As the chapter unfolds, the reader discovers why the most elegant optical ideas are often undone by invisible losses at interfaces, in transport pathways, and under the open sky.

9.1 Quantum-Dot Solar Cells: Band Gaps as Design Variables

The most tempting idea in quantum-dot photovoltaics borrows a familiar pleasure from the lab: make a nanocrystal a little smaller, and its color shifts. Turn that knob deliberately, and a solar-cell designer inherits something close to a superpower. Instead of accepting the band gap as the fixed personality of a material, you treat it as a parameter you can choose, batch by batch, like selecting the tint of a paint.

That change of mindset matters because sunlight is not a single color, but a broad spread from ultraviolet through visible into the near-infrared. A solar cell is, at heart, a selective eater. Photons with too little energy slip through; photons with much more energy are swallowed, but the excess turns into heat. The band gap sets the threshold. Make the gap larger and the cell becomes choosier, ignoring red and infrared light; make it smaller and it can harvest those longer wavelengths that conventional silicon already relies on but many thin-film materials struggle to reach.

Colloidal quantum dots add a twist: the same chemistry that lets you synthesize them as an ink also lets you tune their absorption edge by controlling size. Lead chalcogenide dots, such as PbS and PbSe, became an especially vivid playground because they can be tuned deeply into the near-infrared while remaining processable from solution. That spectral territory is valuable. A conventional solar module is not only a piece of physics; it is also a piece of manufacturing and architecture. Thin, lightweight, roll-coated absorbers that can be engineered to drink in infrared light suggest devices that could be layered, stacked, or integrated in places where thick wafers are awkward. In the most optimistic sketches, one quantum-dot layer might be tuned to harvest the near-infrared that passes through a higher-band-gap cell on top, creating a tandem device that wastes less of the Sun's spectrum as heat.

There is also a subtler point hiding in the word "choose." With bulk semiconductors, changing the band gap usually means changing the material itself: switching from silicon to cadmium telluride, or to a copper-indium-gallium-selenide alloy, or to an engineered perovskite. Each choice drags along a new set of crystal structures, defects, processing temperatures, and supply chains. With quantum dots, composition can stay the same while the absorption threshold shifts. In principle, that makes spectral engineering feel more like device design and less like a periodic-table decision.

A concrete example makes the seduction easier to feel. Imagine a device intended for indoor or low-light environments, where the spectrum is often shifted and the overall intensity is far weaker than direct sunlight. A designer might prefer a larger band gap to achieve a higher open-circuit voltage under dim conditions, because voltage tends to benefit from a generous energy threshold. For a rooftop panel in full sun, a smaller gap may be preferable to maximize current by absorbing more of the red and near-infrared photons that pour down in abundance. With quantum dots, those are not different materials so much as different sizes of the same nanocrystal, potentially fabricated using similar low-temperature steps. The band gap starts to resemble a dial on the workbench.

The dial, however, is attached to a machine full of stubborn gears. A solar cell does not run on absorption alone. It runs on *collected* charges: electrons and holes that must be separated, transported, and extracted at contacts before they reunite and waste their energy as heat or light. The nanoscale freedom that makes quantum dots tunable also makes them surface-dominated. A bulk crystal hides most of its atoms safely inside, but a nanocrystal exposes a large fraction of its atoms at the surface, where bonds are strained or incomplete and where the surrounding ligands and processing chemicals negotiate what electronic states exist. Those surfaces are where the beautiful simplicity of "pick a band gap" meets the less glamorous reality of defects, disorder, and interfaces.

The reversal of the usual "color story" is worth holding onto here, because it explains why the promise is real. Emission is what the eye notices; absorption is what the solar cell lives on. A quantum dot tuned to emit red light has, by the same confinement physics, an absorption edge set in the red or near-infrared. That means it is not merely a tiny lamp; it is a tiny, size-programmable absorber. Solar-cell designers care about the *onset* of absorption because that onset determines which photons contribute to current. So the very knob that makes quantum dots attractive for displays and bioimaging becomes a knob for deciding what slice of sunlight will be turned into electricity.

Yet the key word is “turned.” Absorbing a photon creates an excited state, but a device needs a separated electron and hole that can reach opposite electrodes. In many quantum-dot solar-cell architectures, the dots are packed into a thin film that acts as the light-absorbing layer. That film is not a single crystal. It is more like a densely crowded city made of nanoscale buildings, each building coated in molecular “clothing” left over from synthesis. Those ligands are essential in the flask: they prevent the dots from sticking together irreversibly. But in a solid film, the same ligands can become insulating spacers that make charge transport sluggish. Engineers often swap long, insulating ligands for shorter ones after deposition, or during ink formulation, to bring dots closer and let charges hop between them. Every such swap is both an electrical opportunity and a chemical risk, because it can also expose or create new surface sites that trap charges.

The earliest years of colloidal quantum-dot photovoltaics had the feel of a bold demonstration: if quantum dots can be made and tuned in a beaker, surely they can be painted into a solar cell. In the mid-2000s, efficiencies were well below 1% in many reports, yet the appeal was obvious enough that progress became a recognizable story of rapid improvement. By the early 2010s, carefully engineered devices had moved into the high single digits, and the field began to look less like a curiosity and more like a legitimate member of the photovoltaic family. One sign of that maturation is bureaucratic rather than romantic: the National Renewable Energy Laboratory maintains a “Quantum dot cells” category in its best-research-cell efficiency chart, placing the technology on the same public scoreboard as other emerging solar concepts. That kind of institutional bookkeeping reflects an uncomfortable truth: once a technology is taken seriously, it must be measured by reproducible device performance, not by the elegance of its underlying idea.

Reproducibility brings the band-gap dial back into focus, because dialing size is not the same as dialing a resistor. A solar cell is intolerant of small variations. If the dots in a film have a broad size distribution, then the absorption edge smears out: some dots absorb longer wavelengths, some shorter, and the device behaves as a statistical mixture. That might sound harmless, but it tangles with voltage. In a disordered energy landscape, charges tend to relax into the lowest-energy sites available, which can reduce the achievable voltage relative to what the nominal band gap suggests. A narrow size distribution helps the film behave like a coherent semiconductor rather than a grab bag of slightly different absorbers.

Even with uniform dots, packing matters. A quantum-dot film is a "solid" assembled from particles, and the way those particles touch is not guaranteed to be ideal. Tiny voids, cracks from drying, and grain-like domains can interrupt percolation pathways. A film that absorbs beautifully but conducts poorly will make current only grudgingly. In photovoltaics, that reluctance shows up as low fill factor: the current collapses as soon as the device is asked to deliver power, a symptom that charges are struggling through a resistive, recombination-prone network.

Then there are the interfaces, the places where the quantum-dot layer meets transport layers and electrodes. Solar cells are built from stacks: absorber, electron-selective contact, hole-selective contact, metal electrode, and so on. Each interface is an electronic negotiation. If the energy levels are misaligned, charges pile up. If the interface is chemically reactive, new trap sites appear. If the contact is not selective enough, it can act as a recombination site where electrons and holes meet and annihilate each other right at the exit. In a bulk semiconductor, one may think about doping profiles and junction depths. In a quantum-dot device, the equivalent control often comes from surface treatments, ligand choices, and the subtle chemistry of the interlayers.

This is where the initial fantasy of “spectral matching” needs to be corrected. Traditional solar-cell theory often begins with the idea of an ideal band gap for a single-junction device under sunlight, a value that balances current and voltage. Quantum dots seem to let you pick that value at will. But an efficiently absorbing film with a well-chosen gap can still perform poorly if its surfaces are trap-rich, if its contacts steal carriers, or if transport is slow enough that recombination wins the race.

A useful way to feel the hierarchy of problems is to imagine two devices with identical absorption curves, meaning they harvest the same number of photons. One device extracts charges quickly into well-matched contacts; the other loses many charges to recombination on the way. From the outside, under a lamp, both might look equally dark. Inside, they are not equal at all. The first behaves like a power generator; the second behaves like a sponge that soaks up light and quietly dissipates the energy.

The physics behind these losses is not exotic, just unforgiving. Photo-generated charges are born close together and are naturally attracted. To make electricity, the device must keep them apart long enough to reach opposite electrodes. Any trap site can pin one charge and invite the other to return. Any energetic disorder can shepherd charges into low-energy pockets where they wait too long. Any poorly chosen interface can behave like a cliff they cannot climb. In that landscape, optimizing the band gap is a necessary beginning, not a guarantee of efficiency.

None of this negates the design power of tunability; it simply relocates it. The band gap becomes one variable in a larger design space that includes particle size distribution, surface passivation, inter-dot coupling, film morphology, and contact energetics. The most successful advances have tended to treat those aspects as entangled. Tight control of synthesis improves uniformity and reduces defect densities. Ligand strategies attempt to balance passivation (suppressing traps) against conductivity (allowing charges to move). Interface engineering aims to make extraction fast and selective. The device is not a passive recipient of good material; it is an engineered environment that either honors the nanocrystal's promise or squanders it.

There is also an emotional, human element to this shift from elegance to grind. Many technologies have a phase where a single simple idea captures imaginations. For quantum-dot solar cells, the idea was almost childlike: pick the right size, paint it on, harvest sunlight. The first time a solution-processed quantum-dot film produced a measurable photovoltaic response, it felt like a glimpse of a future where solar energy could be printed rather than grown. The years that followed taught a more mature lesson: the closer a technology gets to usefulness, the more it becomes about the small, careful choices that prevent failure.

Those choices show up in places that are easy to ignore when looking only at spectra. A quantum-dot film is typically made by depositing layers from solution, then performing chemical steps to exchange ligands and densify the solid. Small deviations in those steps can produce large swings in performance. A slightly different drying rate can change film microstructure. A slightly different ligand-exchange completeness can change carrier mobility and trap density. A contact deposited a little too harshly can damage the topmost nanocrystals. In a laboratory, a skilled researcher can coax a device into good behavior. In a factory, the same device must tolerate statistical variation without collapsing into inconsistency. The band gap may be tunable, but a product must also be manufacturable.

The most revealing test of whether band-gap control is functioning as true "design" is whether a device's voltage follows the gap. In an ideal semiconductor, the open-circuit voltage increases as the band gap increases, up to losses that are relatively well understood. In many quantum-dot solids, the voltage falls short of the gap by a distressingly large margin, a signature that recombination and disorder are draining away potential. Improving that gap-to-voltage translation has been a central measure of progress, because it means the device is starting to behave like the textbook object that the tunability argument assumes.

There is a final complication that haunts any discussion of "choose your absorption": sunlight is free, but area is not. A solar cell must absorb strongly enough that the layer can be thin, because thin layers reduce material use and can make charge collection easier. Quantum dots can have strong absorption, but the film must also be dense and electronically coupled enough to behave as a high-quality absorber rather than a dilute paint. The optical promise and the electrical bottleneck meet again: densification often requires ligand removal or exchange, which can increase coupling but also expose surfaces. Device performance becomes a balancing act between letting nanocrystals communicate electronically and keeping their surfaces quiet enough that charges survive.

And yet, despite the obstacles, the central attraction remains unusually durable: in quantum-dot photovoltaics, the absorber's spectral appetite can be engineered without changing the basic manufacturing grammar of solution processing. That is why the technology keeps returning in new forms, from infrared-harvesting layers to tandem concepts to niche applications where lightweight, flexible, or semitransparent devices might matter. The band gap is not a promise of easy efficiency; it is a design lever that becomes powerful only when the rest of the device is built to respect what the nanocrystals can do. When that respect is earned—through uniformity, passivation, and well-behaved interfaces—the idea that began as a color trick starts to look like a practical route for tailoring sunlight into electricity.

9.2 Charge Extraction: The Hard Part Nobody Sees

A quantum-dot solar cell can do the photonic equivalent of catching butterflies all day long and still bring home an empty jar. Absorption is only the first act. The decisive question is whether the excited charges created by that absorbed light can be separated and delivered to the electrodes quickly enough to outrun all the ways they can disappear.

The catch is that a quantum-dot film is not one continuous crystal. It is a solid assembled from tiny, separately grown grains, each wrapped in ligands: molecules that kept the dots stable in a bottle and now sit between neighbors in the film. Those ligands are neither a footnote nor mere packaging. They set the distance charges must hop, they influence how well dots electronically couple to one another, and they help decide whether the dot's surface is calm or riddled with sites where charges vanish.

The problem becomes vivid if you picture what a working solar cell demands of an electron. The electron is created inside a dot, then must move through a dense thicket of other dots, cross at least one engineered interface, and finally enter an electrode. Any step can fail. A long ligand can be a physical spacer that turns a hop into a tunnel through an insulating barrier. A patch of poorly passivated surface can provide a trap state that pins the electron in place. A misaligned contact can make the final handoff energetically awkward, like pushing a suitcase up a stair that turned out to be a wall.

The guiding lesson from years of colloidal quantum-dot photovoltaics is blunt: charge extraction is dominated by interfaces and chemistry. Band offsets (how the energy levels line up), interfacial trap densities, contact materials and architecture, and the chemistry of the quantum-dot solid do more to determine performance than the absorption spectrum alone. The dots may be tuned in the flask, but the current is won or lost at the boundaries between dots and at the boundaries between layers.

That focus on boundaries can feel disappointingly prosaic, as if the grand quantum trick has been reduced to housekeeping. In practice, it is where ingenuity lives. The modern quantum-dot solar cell is as much an exercise in controlling a messy, nanoscale construction site as it is a showcase of tunable band gaps.

A historical anecdote captures how quickly the field learned this. Early devices in the mid-2000s could demonstrate photovoltaic action yet struggled to convert much light into useful power. By 2013, reported efficiencies had climbed to around 8.5% in leading demonstrations, a remarkable jump in less than a decade. That acceleration did not come only from making better dots; it came from treating electrodes, transport layers, and interfacial passivation as first-class design problems. The glamour of tunability brought people in, but the hard-earned gains came from the less photogenic craft of getting charges out.

The ligands provide the clearest example of a double-edged tool. When quantum dots are synthesized colloidally, long organic ligands cling to the surface to prevent aggregation, acting like bristles that keep the particles from touching. In a solar-cell film, that same bristling keeps dots electronically isolated. Charges can still move, but slowly, by hopping. If hopping is too slow, recombination has time to erase the useful charge before it reaches a contact.

So researchers exchange ligands: a chemical step that replaces long molecules with shorter ones or with species that can crosslink neighboring dots into a more conductive solid. This can dramatically improve electronic coupling and carrier mobility. It can also create new problems. Ligand exchange may leave behind under-coordinated surface atoms, or introduce new chemical groups that shift energy levels and disrupt band alignment. The film becomes more conductive but also potentially more trap-rich. A device can gain current and lose voltage, or gain one week and lose the next, depending on how consistently that exchange step is carried out.

That trade is not a flaw in the idea of quantum dots; it is a consequence of working with a material whose defining feature is its surface-to-volume ratio. In a bulk semiconductor, the interior dominates. In a quantum dot, every processing step touches a large fraction of the atoms that matter electrically. The same action that improves transport can change recombination pathways. Device progress often comes from techniques that do both at once: improve coupling while keeping surfaces well passivated and interfaces selective.

There is a phrase used in the field that hints at how radical this can become: the *quantum junction*. In a conventional solar cell, the key rectifying junction is created by doping a semiconductor to make p-type and n-type regions. In quantum-dot photovoltaics, the idea emerged that one could form a junction using quantum-dot layers whose energy levels are tuned (by size and chemistry) to create the needed driving forces for separation and extraction. The promise is elegant: if the dots themselves set the band edges, then band matching across p- and n-type quantum-dot materials could, in principle, be engineered with unusual freedom.

The catch returns immediately: such a junction works only when interfaces are carefully engineered. A rectifying junction is a delicate device. It depends on clean, predictable band offsets and low interfacial trap densities; otherwise the junction turns into a recombination zone rather than a separation engine. The quantum-junction idea therefore throws an unforgiving spotlight on what counts as "clean" in a solution-processed, nanocrystal-assembled solid. It also explains why progress in quantum-dot photovoltaics has required deliberate electrode and transport-layer engineering alongside improvements in nanocrystal passivation and crosslinking. A well-tuned band structure on paper is not enough; the junction must be built in a way that prevents its own interfaces from undoing it.

Everyday manufacturing details become part of the physics here, which is one reason quantum-dot solar cells have such a distinctive personality as a technology. It is not a matter of growing a perfect crystal at high temperature and then etching it. It is a matter of making an ink, depositing a film, swapping ligands, controlling drying and densification, and forming contacts without wrecking what you just assembled. That chain creates a new kind of fragility: the device can be exquisitely sensitive to small variations in steps that, in other industries, might be treated as routine.

Water provides a striking example because it is so ordinary and so easy to overlook. Ambient water can directly change the chemistry of a quantum-dot film and the way it forms during deposition. Humidity can lead to adsorption and hydroxylation at surfaces, altering electronic properties in ways that are not obvious by eye. A film deposited on a dry day can end up measurably different from a film deposited on a humid day, even if the recipe was followed faithfully. The dots are tiny, their surfaces are chemically active, and water is an energetic participant at the nanoscale.

This is not merely a stability story for later outdoor operation; it is a charge-extraction story from the moment of fabrication. Water present during film formation can influence packing density and the microscopic pathways charges must traverse. It can also alter interfacial trap densities by changing which surface groups are present. Under those conditions, a researcher might mistakenly attribute a performance swing to the dots themselves when the real culprit was the invisible moisture content of the laboratory air.

The response has been as practical as it is revealing: deposition techniques that physically manage water. Meniscus-guided coating, for example, has been used to produce dense-packed quantum-dot solids while helping to "extrude" ambient water during deposition. The phrase is wonderfully literal. The coating geometry and controlled evaporation can push water out of the forming film rather than trapping it. The result is not only improved performance but improved thermal stability, a reminder that manufacturing control and charge extraction are entangled. A better film is not just purer; it is assembled under conditions that prevent unwanted guests from being incorporated.

If all this emphasis on extraction feels severe, it helps to connect it to what a solar cell is competing against. Sunlight arrives every day with no invoice, but a solar module must pay its way by being efficient, reliable, and scalable. An elegant absorber that cannot consistently extract charges has no commercial dignity; it is reduced to a lab demonstration. That is why the unglamorous details of contact formation, ligand control, and interface passivation have become the real battleground.

It is also why quantum-dot photovoltaics often advance through improvements that sound modest in isolation. A new ligand-exchange protocol that gives a more uniform coupling across the film. A transport layer that better blocks the wrong carrier, making the contact more selective. An interfacial treatment that reduces trap density at the boundary with a metal oxide. These are the sorts of refinements that do not change the absorption curve dramatically, and they do not make for a colorful photograph. They do, however, change how long a charge survives and how quickly it can be collected, which directly controls current, voltage, and fill factor.

An accessible way to sense the stakes is to think about time. After absorption, an electron and hole exist in a race between extraction and recombination. Extraction is not instantaneous. Hopping from dot to dot takes time; crossing an interface takes time; navigating energetic disorder takes time. Recombination is opportunistic and relentless. If a trap captures one carrier, it creates a waiting room where the other carrier can eventually find it. If an interface is poorly matched, carriers accumulate there, increasing the chance that an electron meets a hole. The device performs well when extraction happens on a timescale that makes recombination improbable.

That race also explains why quantum-dot solar cells can look deceptively good in one metric and disappointing in another. A film can generate a respectable short-circuit current because it absorbs strongly and some fraction of charges make it out. Yet the open-circuit voltage can lag, because recombination through traps and interfaces raises the "dark" currents that erode voltage. Two devices can share a similar absorption edge and still have radically different voltages simply because one has cleaner interfaces and fewer recombination pathways.

The deeper message is that a quantum-dot solar cell is not an optical gadget with electrodes attached. It is an electronic device built from a chemically active, particulate solid. Success depends on arranging that solid so that electronic coupling is strong enough for transport while keeping the surfaces and interfaces quiet enough that charges remain mobile and separate. Ligands, processing conditions, and contacts are therefore not secondary choices; they are the device.

Seen this way, the fascination of quantum-dot photovoltaics shifts from the idea of tunability to the craft of building a functional solid out of nanoscale parts. Each dot begins life as a carefully synthesized object in solution, then becomes a tile in a mosaic where every boundary matters. When the mosaic is assembled with sufficient precision, the charges created by sunlight stop being fleeting excitations and start becoming a steady current that can do work outside the lab.

9.3 Luminescent Solar Concentrators: Windows That Funnel Light

A solar panel announces itself. It is dark, tilted, and unmistakably a machine. A window does the opposite: it is meant to disappear. The ambition behind luminescent solar concentrators is to smuggle energy harvesting into the surfaces that already cover modern life—glazing, skylights, façades, noise barriers along highways—without demanding that every square meter look like a rooftop array.

The trick is optical rather than electrical. Instead of forcing charges to travel through a thick, continuous absorber, a luminescent solar concentrator spreads the work out: it catches light over a large area, converts that light into a new color, and then guides the converted light sideways through a transparent plate to solar cells mounted only along the edges. Geometry does the concentrating. The plate is not merely a support; it becomes a waveguide, a kind of light pipe whose job is to keep photons trapped long enough to reach the perimeter.

The basic sequence is easy to state. Sunlight enters a clear slab, like a piece of acrylic or glass. Within the slab are tiny light-absorbing particles—often colloidal nanocrystals precisely because their absorption and emission can be tuned. They absorb part of the incoming spectrum and then re-emit it at a longer wavelength, as photoluminescence. Because the emitted photons start inside a material with a higher refractive index than air, many of them strike the surfaces at angles that trigger total internal reflection. The photons ricochet inside the slab, traveling sideways until they reach the edge, where a narrow strip of conventional photovoltaic cell converts them to electricity.

This is concentration without mirrors and without tracking the Sun. A roof-mounted concentrator uses lenses or reflective dishes to squeeze light onto a small, expensive solar cell, but it needs optics aligned to the sky. A luminescent concentrator uses the plate itself as the optical element. A building could, in principle, keep its usual vertical windows and still harvest a portion of sunlight, because the concentrator does not require the window to face the Sun directly; it needs only that light enters the plate at all.

The first emotional appeal is architectural: the possibility of power-generating surfaces that remain transparent or at least pleasantly tinted. The second appeal is manufacturing: the luminescent layer can be made by coating or embedding luminophores in polymers and laminates, which resonates with how glass is already treated in industry. The third appeal is that it shifts the most expensive semi-conductor area to the edges. If you can collect over square meters and convert at centimeters of edge-mounted PV, you begin to think in a different economic language.

The hard part is that the device is a photon accountant. At each step in the sequence, losses accumulate, and the accounting is merciless. Some sunlight is never absorbed because the luminophores do not cover the entire spectrum. Some absorbed energy never re-emits as light because the photoluminescence quantum yield is not perfect. Some emitted photons are not trapped by total internal reflection; they escape through the "escape cone," the set of angles that allow light to leave the plate. Some photons that are successfully trapped are lost to scattering or absorption in the host material—waveguide parasitics that grow more severe as the plate gets larger and the path to the edge grows longer.

And then there is reabsorption, the persistent enemy that turns a simple idea into a deep design problem.

Reabsorption (also called self-absorption) happens when a photon emitted by the luminophore is later absorbed by another luminophore before it reaches the edge. Absorption is not inherently bad—after all, the whole purpose is to absorb sunlight—but reabsorption in a concentrator is different. The second absorption is usually followed by a second emission, and each emission offers new opportunities to lose the photon. The photon might escape the waveguide this time. It might be emitted in a direction that takes it away from the edge. It might be emitted slightly redder and begin to drift out of the range that the edge PV converts efficiently. Or the second absorption might be followed by non-radiative loss, turning the photon into heat.

Reabsorption is fundamentally tied to a spectral overlap: if the luminophore absorbs strongly at wavelengths where it also emits, then emitted photons are likely to be re-captured by their neighbors. The obvious solution is to maximize the Stokes shift, the separation between absorption and emission. If emission is shifted far enough to the red that the material no longer absorbs it, then photons can glide to the edges without repeatedly being recycled.

That goal sounds straightforward until you remember the competing requirement: you want the device to absorb a meaningful fraction of sunlight while remaining transparent enough to serve as a window. A luminophore that absorbs only in the ultraviolet could keep the window visibly clear, but it would harvest a limited fraction of solar power. A luminophore that absorbs broadly across the visible can harvest more, but then the window darkens, and the overlap between absorption and emission tends to worsen. The engineer is forced into a three-way negotiation among aesthetics (how much visible transmission, and what tint), energy harvesting (how much sunlight is absorbed), and optical transport (how much of the converted light survives to the edge).

The vocabulary of window design makes this tension concrete. Architects talk about *average visible transmission* (AVT): roughly, the fraction of visible light that passes through. A window with around 50% AVT still feels like a window rather than a shaded panel, especially if the tint is neutral. Energy researchers, meanwhile, ask what efficiency would make such a semitransparent device worth installing at scale, once you account for real window sizes, real building orientations, and the fact that windows are installed for daylighting, not for maximum insolation. Reviews in *Energy & Environmental Science* have framed performance targets in exactly that paired way: any meaningful number must be attached to a transparency and a module scale. A spectacular efficiency in a small, deeply tinted sample does not solve the same problem as a modest efficiency in a large, attractive pane.

Here the tunability of nanocrystal luminophores becomes more than a colorful party trick; it becomes a tool for juggling those constraints. By adjusting their size and composition, one can tune absorption bands and emission peaks. In principle, one can design a luminophore that steals mostly ultraviolet and blue photons (which are more plentiful than people often realize and can be uncomfortable in buildings) and re-emits in a narrow band that is efficiently guided. Or one can engineer a system that collects across multiple spectral regions using multiple luminophores, each chosen to minimize overlap with its own emission.

The phrase "reabsorption-free" appears so often in discussions of practical luminescent concentrators because it is a scaling keyword.

A small plate can tolerate some reabsorption because the average distance to the edge is short. As the device grows to window size, photons must travel longer distances, and even a modest probability of reabsorption becomes a near certainty after many centimeters. What looks like a small spectral overlap becomes a giant loss when the optical path length multiplies it.

A useful mental image is a crowded hallway. If a person must pass through only a few doors to exit, it does not matter much if one door is sticky. If they must pass through a hundred doors, the stickiness becomes the defining feature. Reabsorption is the sticky door of luminescent concentrators.

This is why tandem and stacked architectures have attracted serious attention. Rather than asking one luminophore to do everything, a tandem luminescent concentrator can split the spectrum into layers. One layer absorbs and emits in one region; another layer does the same in a different region. The analogy to multi-junction photovoltaics is deliberate: spectral splitting reduces waste and can, crucially for concentrators, reduce reabsorption by narrowing what each layer needs to absorb and by moving emission into spectral windows where the material is transparent.

A concrete demonstration helps anchor the idea in numbers that feel tangible rather than aspirational. Kaifeng Wu and collaborators, working with Paul Alivisatos's broader scientific lineage that helped bring colloidal nanocrystals into mainstream materials science, reported in 2018 a large-area tandem luminescent concentrator based on engineered nanocrystal emitters. The area exceeded $230\,\text{cm}^2$, large enough to force the device into the regime where scaling losses become unavoidable. The reported optical quantum efficiency was about 6.4% (meaning that a fraction of incident photons were ultimately delivered to the edges), and the solar-to-electrical power conversion efficiency was about 3.1% for the prototype with edge-mounted photovoltaic cells. Those numbers are not meant to compete head-to-head with rooftop silicon; they are meant to show that a window-sized optical device can still deliver a measurable, engineered stream of light to its edges rather than dying of reabsorption and scattering as it grows.

The most interesting aspect of such demonstrations is not any single efficiency figure but the logic of how the device was made to scale. Scaling is often where beautiful lab optics fail. A small sample, illuminated gently and measured carefully, can appear to work wonderfully. When you enlarge it, photons have farther to travel, small imperfections repeat many times, and edge collection becomes a game of survival over distance. Tandem designs, large effective Stokes shifts, and reabsorption-resistant emitters are all responses to that harsh arithmetic.

The other large loss channel, escape-cone loss, is an elegant reminder that luminescent concentrators are constrained by basic optics. Total internal reflection traps light only if it hits the surface at a sufficiently shallow angle. Emission inside the slab occurs in many directions. Some fraction of emitted photons inevitably point into angles that allow them to escape. You can mitigate this with refractive-index engineering, surface coatings, photonic structures, or careful choice of host material, but you cannot wish it away. The device is always balancing how much light is sent into trapped modes against how much leaks out through the faces.

Waveguide parasitics can be surprisingly mundane: absorption by the polymer matrix, scattering from tiny inhomogeneities, and surface roughness that scrambles reflections. A plate that is optically perfect for a few centimeters may become effectively lossy over a meter. That is why "finished glazing" is a different achievement from a lab demonstration. A building window must meet safety standards, durability requirements, and long-term optical clarity. It must be laminated, sealed, and cleaned. It must survive temperature swings and sunlight without yellowing or hazing. Some recent work has begun to speak in that language, including a preprint report describing a certification-grade luminescent concentrator photovoltaic glazing prototype. Preprints are not verdicts, but their existence signals that the technology is being pulled toward the world of architects, building codes, and procurement rather than staying only in the comfort of benchtop optics.

The real world introduces loss channels that feel almost insulting in their simplicity. A device can be foiled by dust. It can be foiled by water droplets. It can be foiled by the dried residue left behind after water evaporates.

Studies of environmental effects on luminescent concentrators have shown that where contaminants land matters. Dust on the outside surface mostly shades the device, reducing input light in the obvious way. Dust on the inside surface can, counterintuitively, increase output in some cases by backscattering transmitted sunlight into angles that the plate can trap and guide. Water droplets, on the other hand, can be especially detrimental because they create local optical interfaces that disrupt total internal reflection, giving trapped light an exit route. The worst case is often dried deposits: they combine shading with waveguide losses, turning a smooth internal reflector into a rough, light-leaking boundary. These are not exotic degradation mechanisms. They are the same annoyances that make any window grimy—only here they translate directly into lost electrical power.

That everyday vulnerability loops back to why luminescent concentrators are so appealing: they invite solar energy into the places where people live and work, where surfaces are touched, cleaned, exposed to rain, and judged aesthetically. A rooftop panel is allowed to look utilitarian. A lobby's glass wall is not. The device must satisfy human expectations as well as optical ones.

There is a quiet beauty in the way the physics meets the design brief. The technology takes something that solar engineers usually treat as a nuisance—photoluminescence, the re-emission of absorbed light—and turns it into a transport mechanism. It uses total internal reflection, a phenomenon most people first meet in the sparkle of a diamond or the illusion of depth in a swimming pool, and recruits it for energy harvesting. It asks nanocrystal chemists not merely to maximize brightness but to sculpt spectra to evade reabsorption. It asks materials engineers to treat a pane of "glass" as a high-performance optical component whose defects are not cosmetic but energetic.

If such windows become common, their success will not be announced by a dramatic new skyline of shining panels. It will be more subtle: edges that quietly warm with guided light, narrow photovoltaic strips doing their work out of sight, and buildings that harvest a fraction of the Sun simply by being buildings.

9.4 Durability Under the Sky

A solar device has an easy life in a journal article. It is measured fresh, on a clean substrate, under a calibrated lamp, with the humidity and temperature of the laboratory politely held at bay. Outside, nothing is polite. Sunlight arrives with ultraviolet in it, rain arrives with dissolved gases and dirt, temperatures swing from winter nights to summer afternoons, and the device spends its working life under electrical bias, forever pushing charges toward contacts that would sometimes prefer to undo the separation.

For quantum-dot energy devices, that outdoor reality is not a minor footnote. Durability is governed by coupled stressors: oxygen and moisture, illumination (including UV), heat, and operating bias.

The word *coupled* is doing real work. These stresses do not take turns. They overlap and amplify each other. Heat makes chemical reactions faster. Light creates excited states that can open reaction pathways that are dormant in the dark. Electrical bias can drive ions and charged defects to migrate and can change which interfacial reactions are favored. Moisture and oxygen are not simply contaminants; they are chemically active participants, able to reshape surfaces and interfaces that were carefully tuned during fabrication.

It helps to be concrete about what is vulnerable. A quantum-dot photovoltaic film is a solid assembled from particles. Its performance depends on the integrity of each dot's surface and on the quality of the inter-dot connections created by ligand exchange and processing. Those are precisely the places where the environment can interfere. Oxygen can oxidize reactive components. Water can adsorb, alter bonding, and change the local dielectric environment that charges experience. UV light can break chemical bonds in ligands and polymers, creating new trap-rich sites and changing how well surfaces are passivated. Thermal cycling can create mechanical stress in layered stacks, forming microcracks or delaminations that let more oxygen and water in. A device can look perfectly intact and still be electronically wounded at the nanoscale.

The same is true, in a different way, for luminescent solar concentrators. Their central job is optical: absorb, re-emit, and guide light to the edges. That guiding depends on total internal reflection inside a transparent plate. A single water droplet on the surface can create an optical interface that steals guided photons. A dried residue can roughen the surface and scatter light out of trapped modes. Dust can shade; dust can also, depending on where it lands, backscatter light in surprising ways. These are failures that do not require any chemical decomposition at all. The device can be chemically stable and still lose performance because its optical pathways are disrupted by ordinary grime.

Durability, then, is not a single problem. It is an argument between elegant nanoscale design and everything that happens to a surface left outdoors.

There is a quiet but crucial distinction between two kinds of progress in solar research. One is measured by efficiency records. The National Renewable Energy Laboratory's efficiency chart exists for a reason: it gives the community a shared, independently curated scoreboard for what can be achieved under controlled conditions. The other kind of progress is measured by whether a device can keep working after months of heat, light, and humidity, with its edges sealed imperfectly and its materials slowly reacting to their surroundings. High efficiency does not imply long outdoor lifetime. The two metrics can even be in tension, because some routes to higher performance involve more aggressive processing or thinner layers that can reduce robustness.

Quantum-dot photovoltaics have felt this tension sharply. The devices that perform best often rely on carefully engineered interfaces and highly coupled films to extract charges efficiently. Those same interfaces and films can be chemically delicate. A ligand that is optimal for electronic coupling may be less resistant to oxidation or hydrolysis. A transport layer that gives excellent band alignment might be more reactive with the quantum-dot layer under illumination. A thin device stack might be more sensitive to pinholes or to the slow diffusion of water vapor.

So what actually happens when these devices sit under the sky?

Start with oxygen. Many high-performing colloidal quantum-dot solar cells use lead chalcogenide absorbers, and chalcogenides can be susceptible to oxidation. Oxidation changes the electronic structure at surfaces and interfaces: it can create new defect sites, shift energy levels, and modify how charges move. Oxygen can also react with ligands and with organic components in the device stack, generating byproducts that were never part of the design. Under illumination, these reactions can accelerate. Light creates excited carriers, and those carriers can participate in redox chemistry that does not occur in the dark.

Now add moisture. Water is chemically and physically disruptive. It can adsorb onto surfaces and change which chemical groups are present. It can facilitate hydrolysis reactions that cut ligands or alter bonding at interfaces. It can plasticize polymers and change mechanical properties. It can condense and evaporate repeatedly, hauling dissolved salts and pollutants into crevices and leaving deposits behind. In layered devices, water can diffuse along interfaces in ways that are hard to detect until performance has already shifted.

Heat and thermal cycling add yet another dimension. Any real module will experience daily temperature swings. Materials expand and contract at different rates, and those mismatches generate stress. Over many cycles, that stress can create microcracks, open pathways for ingress, and degrade contact integrity. Heat also accelerates diffusion and chemical reactions, turning a slow degradation pathway into a rapid one during hot spells.

Electrical bias may be the least intuitive stressor for non-specialists, yet it matters because a working photovoltaic device is not a passive object. It is a system with built-in electric fields and steady currents under sunlight. Those fields can drive charged species to drift. Interfaces that are stable at open circuit can behave differently under operating conditions. Bias can also influence which chemical reactions are favorable at contacts. A device that seems stable in storage can age quickly under real operation, precisely because operating conditions activate processes that idle samples never experience.

These stressors interact. Moisture can make oxidation easier. Heat can make moisture diffusion faster. Light can make oxidation chemically more aggressive. Bias can change the local charge distribution and thus change which surface sites are reactive. That is why durability tests that isolate a single variable can be misleading: they are necessary for diagnosis, but field performance often reflects the combined effect.

At this point the discussion can feel abstract, so it is worth returning to an example where a very ordinary variable—ambient water during fabrication—bridges manufacturing and lifetime.

The way a quantum-dot film is deposited can seed its long-term instability. Ambient water affects film formation and chemistry. A film cast in humid air can incorporate water at interfaces or alter surface groups in a way that changes its electronic properties immediately and its degradation pathway later. The striking part is that this is not merely a matter of storing devices in dry boxes afterward. The initial microstructure and chemical state of the film, set during deposition, can determine how easily oxygen and moisture attack later and how well the device tolerates heat.

This is why process choices that manage water residuals have drawn attention. Meniscus-guided coating, for example, can produce dense-packed quantum-dot solids while reducing residual water by physically pushing it out during the controlled deposition. The phrase from the literature is that the method can "extrude" ambient water. That manufacturing tweak can improve performance and thermal stability together, because it changes both the electronic coupling and the chemical starting point of the film. It is a neat illustration of a broader principle: durability is often written into a device at the moment it is made, not only during its later exposure.

Encapsulation is the obvious response, and for good reason. Encapsulation means surrounding the device with barrier layers that slow the ingress of oxygen and water, while also protecting against mechanical damage and UV exposure. In photovoltaics generally, encapsulation is what turns a lab device into a module. For nanocrystal-based devices, it can be enabling rather than merely protective, because the active layers are so surface-sensitive.

But encapsulation is not magic. Barrier layers are never perfectly impermeable, and edges are hard to seal. Water vapor and oxygen find pathways along interfaces, through pinholes, and around contacts. Encapsulation can also create its own problems if the materials used react with the device stack or if the sealant outgasses chemicals that alter surfaces. A truly durable design treats encapsulation as part of the system: barrier performance, adhesion, UV stability, and mechanical compatibility all matter.

The story becomes even more vivid for luminescent solar concentrators, where the environment can attack not only the chemistry but the optics.

A luminescent concentrator relies on total internal reflection to guide emitted photons to the edges. Total internal reflection is a surface phenomenon; it depends on having a clean, smooth interface between the plate and the surrounding air (or between layers in a laminate). Anything that changes the surface refractive index locally, or roughens the interface, can turn guided photons into escaping photons.

Everyday environmental interactions can directly alter waveguiding. Water droplets degrade waveguiding by forming a new optical interface at the surface. A droplet is not just water; it is a lens-like perturbation with a refractive index closer to the plate than air is, opening an escape channel for photons that would otherwise be reflected internally. Dried deposits are often worse because they create both shading and waveguide losses: they block incoming light and also scatter guided light out of the plate. Dust effects can depend on whether deposition is on the outer or inner surface. Dust on the outside mostly shades; dust on the inside can sometimes increase output by backscattering transmitted sunlight, a counterintuitive reminder that waveguides are sensitive to where scattering occurs.

This is a particularly human kind of fragility. It connects photovoltaic performance to cleaning schedules, to rainfall patterns, to the chemistry of hard water, to urban pollution, to whether a building's glazing is exposed to wind-blown dust or sheltered under an overhang. It also suggests a design space that is not purely about emitters and spectra. Surface coatings that repel water, textures that discourage residue formation, and lamination strategies that protect the waveguiding surfaces can matter as much as photoluminescence quantum yield.

Durability discussions can easily become pessimistic, so it is worth saying what has improved. Researchers now treat stability as a multistressor engineering problem, not as a single material weakness. In quantum-dot photovoltaics, stability strategies span materials, interfaces, and encapsulation. Materials approaches include more robust passivation schemes and the use of core–shell nanocrystals in some contexts to protect sensitive cores. Interface approaches include choosing transport layers and contact materials that are less reactive and designing interlayers that block unwanted chemical exchange. Encapsulation approaches include barrier films and edge seals borrowed and adapted from the photovoltaic industry and from packaging for organic electronics.

In luminescent concentrators, materials strategies include choosing hosts that resist yellowing under UV, emitters with minimal photobleaching, and spectral designs that reduce reabsorption so photons do not undergo repeated absorption–emission cycles that multiply the opportunities for loss. Device strategies include geometries that shorten average photon path lengths and edge-coupling designs that collect light more effectively. And, increasingly, there is attention to the standards and practicalities of glazing: safety glass requirements, lamination processes, thermal expansion compatibility, and the long list of tests any building material must pass.

Claims of "commercial readiness" should be tethered to standards and verified performance. Some recent reports are moving toward certification-grade luminescent concentrator glazing, but they may appear first as preprints and should be treated as forward-looking rather than definitive. This is not cynicism; it is simply respect for how difficult it is to make a device that performs in the field for years and can be installed in buildings with the same confidence as ordinary glass.

A final anecdote, familiar to anyone who has watched a promising technology mature, is that the most difficult problems often stop being the ones that first attracted attention. Early on, the headline challenge for nanocrystal energy devices was achieving respectable efficiency. That challenge made sense when devices were fragile and underperforming. As efficiencies improved, a more sobering truth emerged: a device that works only when freshly made is not an energy technology. It is a measurement.

Under the sky, sunlight is not only fuel. It is also a stress test. Heat is not only a background condition. It is an accelerator. Water is not only a threat of corrosion. It is an optical spoiler and a chemical participant. What determines success is not a single clever material but a whole chain of choices: how the film was formed, how interfaces were engineered, what barriers were applied, how edges were sealed, and how the device responds to the ordinary insults of weather. The reward, if those choices cohere, is a class of solar devices whose most delicate features—nanoscale surfaces and tunable spectra—survive long enough to matter in the places where energy is used.

Chapter 10

The Price of Brilliance: Toxicity, Rules, and Better Materials

The same nanoscale features that make quantum dots so useful also make them hard to judge by simple labels like safe or dangerous. Beneath every vivid display pixel and every biomedical probe lies a more complicated story—of chemistry escaping its container, rules quietly redesigning entire industries, and a race to invent materials whose brilliance does not come with so heavy a moral bill.

10.1 What Makes a Dot Dangerous? Composition, Pathways, Context

A question that comes up quickly, in labs and in living rooms, is blunt: "Does it contain cadmium? Does it contain lead?" The question is not wrong. Cadmium and lead have long histories as poisons, and a material that can release them deserves scrutiny. Yet the question is also incomplete in a way that matters for real lives and real policy. A nanocrystal sealed inside a display, locked under layers of plastic and glass, is a different object—in practical consequence—from the same nanocrystal as a dry powder that can become airborne during manufacturing, or from that nanocrystal after years of sunlight, heat, and oxygen have roughened its protective coatings, or from its ash after incineration.

The more scientific question is therefore conditional: under what circumstances can a harmful species be released, taken up by a body or an ecosystem, and do damage?

That conditional wording may feel like hedging. It is the opposite. It is an invitation to separate two ideas that are often mashed together in public debate.

Hazard is what a substance can do, in principle: the capacity to harm. Cadmium ions, for example, can interfere with essential proteins and are toxic to kidneys and other organs at sufficient doses. Tiny particles can irritate lungs if inhaled. Certain surfaces can drive chemical reactions that generate reactive oxygen species, a catch-all term for aggressive forms of oxygen chemistry that can stress cells.

Risk is what a substance is likely to do in a particular setting: hazard *combined* with exposure. Risk depends on how much gets into a body, by what route, for how long, and in what chemical form. A sealed vial of chlorine gas is hazardous; it is not, by itself, a high risk to a person who never opens it. A small leak changes that instantly. The same logic holds for glowing nanocrystals, except that leaks can be slow, subtle, and shaped by time and environment.

With quantum dots, hazard and risk are separated by engineered barriers that are both chemical and physical. The dot's core is often a semiconductor made from elements that would raise eyebrows on a chemical inventory: cadmium selenide, cadmium telluride, lead sulfide. Around that core, designers commonly add a shell, frequently a more stable semiconductor such as zinc sulfide, and then attach organic molecules called *ligands* that bind to the surface and help the dots stay dispersed in a liquid or compatible with a polymer. Shells and ligands are not decorative. They are containment, stability, and interface all at once. When they do their job, they reduce the dot's chemical reactivity and slow the escape of metal ions. When they fail, the dot's most sensitive parts—its surface and its core—are suddenly in contact with the outside world.

Failure has many faces. Heat can loosen ligands or soften the polymers that hold dots in place. Light can trigger photochemical reactions that create oxidizing conditions right at the surface. Oxygen and water can creep in over years and gradually corrode protective layers. Acidic environments can dissolve shells that were stable at neutral pH. Mechanical wear can create fresh surfaces and fine debris. None of these processes needs to be dramatic to matter; a slow change in the chemistry of a surface can change how particles clump, how they travel through water, and whether living cells absorb them.

It helps to imagine the dot not as a bead of one material but as a layered object with a personality that shifts with its surroundings. A pristine, well-passivated core-shell dot in a bottle is one character: bright, stable, relatively well-behaved. The same dot, after months of ultraviolet light and oxygen, can become another character: less emissive, more chemically reactive, and more likely to release ions or fragments. Toxicology studies have repeatedly found this kind of context dependence: "fresh" particles can look tolerable under certain tests, while oxidatively degraded particles can provoke stronger biological responses. That is not a paradox; it is a reminder that the environment continues the chemistry long after synthesis.

Surface chemistry sits at the center of this story. Ligands are small, but they control big outcomes. A ligand can make a particle effectively water-loving or water-hating; it can give the surface a positive or negative charge; it can determine whether proteins stick to the surface in the bloodstream, forming what researchers call a *protein corona*. Those choices influence where particles go in a body and how long they remain there. In practice, two samples with the same core composition can have very different biological fates because the body does not encounter “cadmium selenide” in the abstract; it encounters a nanoscale object with a particular surface that invites or resists interactions with cells, membranes, and immune systems.

The route into the body is equally decisive. Inhalation, ingestion, dermal contact, and injection are not interchangeable. The lungs are exquisitely vulnerable to fine particles because the surface area is vast and the barriers are thin. The gut is chemically harsh but also designed to keep many things out; much that is swallowed passes through without entering the bloodstream, though not everything. Skin is an excellent barrier for many substances, yet cuts and compromised skin change the calculation. Injection bypasses the body’s outer defenses entirely and places materials directly into a fluid environment where proteins, salts, and immune cells begin interacting immediately.

These routes map onto the places where quantum dots are made and used. The same nanocrystal can be safe enough in one context and a meaningful hazard in another, simply because the likely exposure changes.

Consider the manufacturing bench, where some of the most celebrated dots were first perfected. The classic colloidal synthesis methods that made the field explode in the 1990s relied on hot, reactive chemicals, careful exclusion of oxygen and water, and the handling of cadmium precursors that no chemist would treat casually. Early labs were not thinking only about brightness; they were learning how to keep air and moisture from ruining a reaction, how to purify products, how to control size. Safety practices were often those of synthetic chemistry: gloves, fume hoods, sealed glassware, waste containers. As quantum dots moved from small research batches to industrial quantities, the emphasis shifted from artisanal skill to reproducible processes, and with that shift came a different exposure profile: more workers, more material, more opportunity for spills, aerosols, and waste.

A practical industrial problem makes the risk pathway vivid. Quantum dots are often processed as dispersions—tiny particles in a solvent—because dispersions can be coated as thin films, printed as inks, or mixed into polymers. In a liquid, inhalation risk is lower than for a dry powder. But drying steps exist everywhere in manufacturing: cleaning filters, evaporating solvents, spray-coating, curing films. Wherever a wet dispersion becomes a dry residue, particles can become airborne if the residue is disturbed. Occupational exposure is therefore not a philosophical concern; it is a matter of mundane tasks like scraping a container, cleaning a reactor, or emptying a vacuum line trap. The hazard may be in the core composition, but the risk is in the workflow.

Now move to the consumer product, where the most familiar use is in displays. Here, the dots are rarely loose. They are embedded in polymer films or in encapsulated components designed to sit behind a screen, converting blue light into narrower bands of red and green. Good device engineering keeps them away from fingertips, mouths, and household air. Under normal use, the likely exposure is minimal, which pushes risk down even if the dot's intrinsic hazards are nontrivial.

Yet "normal use" is a term with edges. Devices heat up. Sunlight streams through windows. Screens are dropped. Plastics age. A well-designed film can still be vulnerable to the slow stressors of time: oxidation, light exposure, and thermal cycling. If the polymer matrix cracks or yellows, if oxygen permeability increases, if the dot's ligand layer is slowly stripped or rearranged, the chemistry can begin to drift. The most immediate consequence is optical: the picture degrades. The deeper consequence is that the barriers that separate hazard from exposure become less trustworthy.

It is tempting, at this point, to ask for a binary verdict: safe or unsafe. Real materials seldom grant that comfort. The more honest question is: safe *for whom*, under what conditions, and with what safeguards?

Medical use turns that question into a moral demand, because the exposure route is intentionally intimate. When a dot is injected as an imaging label, the design goal is not containment inside a plastic film; it is navigation inside a body. The dot must remain dispersed in water-based fluids, avoid immediate clearance, and sometimes bind to specific biological targets. Those demands push surface chemistry into difficult terrain. Water solubility often requires swapping hydrophobic ligands for hydrophilic ones or wrapping the dot in polymers, proteins, or silica-like shells. Each modification changes how immune cells respond and where the dot accumulates.

A recurring observation in animal studies is that many nanoparticles introduced into the bloodstream are taken up by the liver and spleen, organs that act as filters and immune sentinels. Persistence there over weeks is not surprising; it is a sign that the body is treating the particles as foreign objects to be sequestered. Whether that persistence is acceptable depends on dose, on whether the particles degrade and release ions, and on what biological effects follow. Again, the core composition matters, but it is not the whole story. Surface coatings can lower immediate toxicity by reducing ion leakage and reactivity, yet they can also create long-lived materials that the body struggles to clear. A coating that makes a dot *less* chemically aggressive can make it *more* persistent, and persistence is its own kind of risk when long-term effects are uncertain.

There is a historical echo here in the broader story of biomedical materials. Early enthusiasm for "inert" implants and particles often collided with the reality that the body is not a passive container. It reacts, walls off, inflames, remodels. In the case of quantum dots, the early 2000s saw an explosion of papers demonstrating dazzling cellular images—dots tracking receptors, labeling tumors in mice, surviving photobleaching better than many dyes. That brightness was real. But so were the questions that followed: Where do the dots go after the picture is taken? Do they stay intact? Do they break down under biological conditions? Does a bright signal today imply a quiet presence tomorrow? The field's maturation has involved learning to ask these questions as part of the design brief, not as an afterthought.

At the middle of all these contexts sits a deceptively simple idea: a dot's "danger" is a chain, and the chain has links that can be strengthened or weakened.

Thinking this way changes how familiar debates look. Take cadmium again. A cadmium-based dot has a high intrinsic hazard *if* cadmium becomes bioavailable as ions or reactive fragments. A robust shell can make that conditional hard to satisfy in normal device operation. But shells are not diamond; they are semiconductors with their own defect chemistry. Under oxidative or photolytic stress, shells can develop cracks or dissolve, and ligands can detach. The very features engineered for brightness—a clean interface, a passivated surface—are also features that determine whether the core stays sealed.

This is why toxicology for nanomaterials can feel frustratingly inconsistent to outsiders. One study finds little effect at a given dose; another reports inflammation or oxidative stress. Sometimes the difference is the species or the cell line. Often it is something more prosaic: one material was freshly prepared and well-coated, another was aged or intentionally degraded; one was dispersed with a stabilizing polymer, another clumped; one test medium stripped ligands, another did not. In ordinary chemicals, "purity" is mainly about how much of a single molecule you have. For nanocrystals, "purity" includes the integrity of shells, the identity of ligands, and the particle's tendency to aggregate. If those are not characterized carefully, comparisons become muddy.

A concrete example makes the chain-of-contingencies idea less abstract. Imagine a factory producing quantum-dot films for displays. The dots arrive as a liquid concentrate, are mixed into a polymer, coated onto a substrate, and laminated into an optical stack. During coating, the solvent evaporates, and workers may deal with vapors and, in some processes, fine droplets. The hazard of the core composition is present, but the most likely near-term risk is inhalation of aerosols or skin contact with dispersions. Good ventilation and protective equipment are not bureaucratic extras; they interrupt the exposure link in the chain.

Now imagine the film ten years later, when a television is discarded. If the unit is recycled responsibly, the film may remain encapsulated and be separated as a component. If it is shredded indiscriminately, fragments of polymer containing dots can become part of a dust stream. In a landfill, slow leaching by water percolating through waste can expose the film to acids and oxidants that were never part of the product's design environment. In an incinerator or an open fire, high temperature and oxidative conditions can destroy organic ligands and polymers, potentially leaving inorganic residues that behave very differently from the original dot-in-plastic. The same nanocrystal, in other words, can present different risks at different life stages even when its starting composition never changes.

Dose threads through every one of these scenarios. A hazard without a meaningful dose is often a theoretical worry; a modest hazard at high dose can become a real emergency. Dose is not only "how much" but also "how available." A gram of cadmium locked in a stable lattice and sealed away is not equivalent to a milligram of cadmium ions dissolved in water, even though the element is the same. Bioavailability is why coatings matter so much: shells and ligands can turn a potentially toxic element into something closer to a trapped guest.

The tricky part is that trapping is rarely permanent. Time is a reagent. Sunlight is a reagent. Mechanical stress is a reagent. That is why life-cycle thinking is not a moral slogan but a scientific necessity. A dot that is safe in a product can still be a concern in a waste stream; a dot that is tolerable in a controlled clinical dose can still be problematic if it persists, accumulates, or degrades into more bioavailable forms.

The most uncomfortable implication is also the most useful: there is no single toxicity label that can be stamped onto "quantum dots" as a category. The relevant object is the *particular* dot, with its core, shell, ligands, and packaging, moving through a *particular* set of environments, meeting *particular* exposure routes. The same scientific mindset that made nanocrystals bright—attention to surfaces, to defects, to interfaces—is what makes their safety legible. Safety is not only about what a dot *is*; it is about what the dot can *become* when the world does what the world always does: heats, shines, oxidizes, abrades, and carries materials into places their inventors never intended.

10.2 Regulations as Invisible Engineers

A modern phone or television comes with a thin booklet of legal language that almost nobody reads. The screen itself comes with no such booklet at all. Yet behind that unremarkable rectangle sits an entire architecture of rules: restrictions on certain elements, reporting obligations for unusual material forms, standardized methods for measuring what a material *is*, and approval pathways that determine what counts as an acceptable medical product. None of these rules emits light, but they decide which light-emitting materials are worth perfecting.

Engineers often talk as if invention happens first and regulation arrives later, like a referee entering after the game begins. In practice, rules quietly set the shape of the playing field. They determine which research programmes look investable, which supply chains can survive scrutiny, and which technical compromises are rewarded rather than punished. For nanocrystals that are dazzling in the lab but ambiguous to a non-expert eye, regulation becomes a kind of invisible engineer: it nudges chemists toward one core composition rather than another, pushes device makers toward thicker encapsulation, forces manufacturers to treat reproducibility as a moral and commercial necessity, and turns measurement into a prerequisite for being allowed to sell anything at all.

Two worlds show this especially clearly. Consumer electronics faces substance restrictions that were never written with nanocrystals in mind but now shape them. Medicine faces product-specific evaluation that is less about banning elements and more about proving, with exhausting thoroughness, what happens in a body over time. The same bright particle can look like a manageable compliance task in a television and a daunting uncertainty in an injectable imaging agent.

Start with the European Union's Restriction of Hazardous Substances directive, usually referred to as RoHS. RoHS is simple in spirit: certain hazardous substances should not be present above set thresholds in electrical and electronic equipment, unless there is a justified exemption. Simple rules have sharp teeth. They do not ask whether a particular company is careful; they ask what the product *contains*, as defined by law. That legal framing is powerful because it reaches far beyond the factory floor. It shapes procurement, documentation, customer requirements, and the internal arguments inside firms about which materials are worth the trouble.

Quantum-dot displays collided with RoHS in a way that became a cautionary tale and an innovation driver at the same time. Cadmium-based nanocrystals were, for years, the easiest route to narrow, saturated reds and greens. Their optical performance helped make a generation of high-colour-gamut screens possible. Yet cadmium is also one of the substances RoHS restricts. The result was not a simple ban and not a simple free pass. It was something more characteristic of modern regulation: a tightly defined exemption category, with conditions and time limits, sitting inside a larger restriction framework.

The details matter because the details are where engineering happens. RoHS does not only list "cadmium is restricted"; it also contains an exemption system with limited validity periods and renewal logic. When a technology is new and alternatives are immature, exemptions can function as a bridge to market. When alternatives improve, exemptions can be allowed to expire. The bridge is never neutral. It creates an incentive to build the alternative before the clock runs out.

A concrete illustration arrived with Commission Delegated Directive (EU) 2024/1416, which grants a specific exemption for cadmium in downshifting nanocrystals directly deposited on LED semiconductor chips for display and projection applications. This kind of language is unmistakably legal, but it is also unmistakably technical. The exemption is not for "any use of cadmium nanocrystals"; it is for a particular implementation, described in scope terms that hinge on how the nanocrystals are integrated into a device. That scope wording is a design constraint. If an approach does not match the defined category, the exemption does not apply, regardless of how safe a company believes its packaging to be.

The same directive signals a substitution horizon, noting that alternatives may be available by the end of 2027. That date works like a distant alarm clock. It changes what research managers green-light, what suppliers develop, and what customers demand in contracts. It also changes the strategy of incremental improvement. A company deciding whether to invest in refining cadmium-based materials or in pushing cadmium-free replacements must weigh not only performance and cost, but also whether the legal permission might evaporate on a specified timetable. This is regulation acting as a scheduling device for innovation.

RoHS also shows how law defines technology categories. RoHS Annex III point 39 (as amended in consolidated texts) describes certain quantum-dot applications in legal scope terms. That may sound bureaucratic, but it has a subtle effect: it tells the market what a regulator thinks the technology *is*. The definition shapes compliance claims, third-party testing, and disputes about borderline cases. It can also freeze a moment in technological history. If a new device architecture emerges that does the same optical job in a different physical way, the firm may find itself arguing that the new approach fits inside old legal wording—or discovering that it does not.

For a general reader, the striking point is that this legal machinery reaches into the lab. A chemist might dream up a gorgeous new cadmium-based nanocrystal that solves a stability problem. A product team might still decline it, not because it fails technically, but because it increases compliance risk, complicates documentation, or places the product under an exemption that could be withdrawn. The chemical periodic table becomes, indirectly, a business plan.

A second regulatory force is quieter but just as influential: the demand to characterize materials in ways that make evidence portable across borders and credible to non-specialists. Here the European Union's REACH framework provides a revealing example. Under REACH nanoforms requirements (Regulation (EU) 2018/1881), nanoform characterization and risk assessment are explicit legal obligations. This moves "what exactly is your material?" from a scientific curiosity to a market-access requirement. Companies must specify size distribution and other defining properties of the nanoform, and they must connect those properties to safety claims.

That requirement changes innovation in a practical way. In a research context, it is common to tolerate some variability: a batch that is slightly broader in size distribution, a ligand mixture that is not perfectly identified, a product that is "close enough" for a proof-of-concept plot. In a regulated commercial context, that fuzziness becomes expensive. If the legal entity is a nanoform defined by certain characteristics, then drifting away from those characteristics can force new testing, new documentation, or a re-interpretation of what is being sold. Measurement becomes part of manufacturing, not an afterthought.

This is where the phrase "paperwork becomes physics" earns its keep. When regulators ask for a defined nanoform, a firm must invest in metrology: tools and methods for measuring size distributions, surface properties, dissolution behaviour, and other parameters that determine both function and safety. Those measurements then feed back into process control. A synthesis route that produces spectacular brightness but inconsistent size may lose to a slightly duller route that produces tight distributions batch after batch, because the latter is easier to defend in a dossier and easier to reproduce across factories.

The United States offers a parallel story through a different legal tradition. Under TSCA, the U.S. Environmental Protection Agency treats many nanoscale materials as chemical substances and uses reporting, information gathering, and premanufacture review as key control points. This shapes commercialization upstream. The practical message to a company is clear: bringing a new nanoscale substance to market is not only an engineering challenge; it is also a data-generation challenge, and it begins before high-volume production. Firms build regulatory work into their development timeline, which can steer them toward materials with known profiles or toward modifications that reduce uncertainty.

Standards bodies add another layer of invisible engineering. Regulations often say, in effect, "provide reliable evidence." Standards help decide what reliability looks like. Measurement and characterization standardization function as governance infrastructure because they determine what counts as comparable data. OECD TG 125 and ISO/TR 13014 are examples of this kind of infrastructure: they offer ways to describe and measure nanoscale materials so that results can be interpreted and accepted across institutions.

This matters because nanomaterials make it easy to fool oneself. A particle's behaviour depends on aggregation, on impurities, on how it is dispersed, on the chemistry of the medium. Two laboratories can test "the same" substance and obtain different answers because their materials were not equivalent in the ways that matter. Standardized characterization does not eliminate uncertainty, but it reduces a particular kind of confusion: uncertainty caused by not knowing what was actually tested.

There is a human story behind this, too, drawn from the culture of industrial quality control. The electronics industry learned long ago that a component is only as good as its specifications and traceability. A manufacturer does not buy "copper"; it buys copper with a specified purity, form, and supplier history. The move from artisanal chemistry to high-volume production forces nanocrystals into that world. When a standards document insists that size distribution be measured and reported in a particular way, it is not merely satisfying a regulator. It is making nanocrystals legible to a supply chain that is built on legibility.

That supply-chain legibility becomes especially sharp when a restricted substance is involved. If a product contains cadmium-based nanocrystals under a permitted exemption, every downstream actor may ask for proof: proof that the material fits the exemption category, proof that the cadmium content is within thresholds, proof that the supplier's process is consistent. This creates a kind of compliance gravity. The legal risk does not stay with the chemical supplier; it is shared, often uneasily, by brands that sell finished devices and fear recalls or reputational damage. The result is that some companies will choose a lower-performing material that avoids restricted substances, simply to simplify the compliance burden across the chain.

The regulatory landscape changes again in medicine, where the aim is not to keep a material out of the body but to justify putting it in. Here, blanket substance restrictions are not the primary tool. Instead, product-specific evaluation dominates. FDA guidance for biomedical nanomaterials emphasizes characterization and manufacturing controls, and it pushes developers to treat nanomaterial identity and reproducibility as central to safety and efficacy. The regulator is not only asking "Is this ingredient hazardous?" but also "Is this product consistent? Does it behave in the body the way you claim? Can you manufacture it the same way next month and next year?"

This kind of oversight reshapes design choices early. A research team might prefer a surface coating that yields the brightest signal in a mouse study. A development team, thinking about eventual clinical use, might prefer a coating that is less bright but easier to manufacture reproducibly, easier to sterilize, and easier to characterize. A ligand that is a convenient mixture in a research lab can become a nightmare in a regulated product if the mixture varies between batches. A complex multi-step coating may look elegant scientifically and still be rejected because it introduces too many variables to control at scale.

Medical regulation also turns pharmacokinetics—what the body does to a substance over time—into an engineering target. A nanocrystal that persists for weeks in the liver might be acceptable in a narrowly justified context and unacceptable in another, depending on benefit, dose, and alternatives. That judgement is not made in an abstract debate about whether an element is "toxic." It is made through a product dossier that ties material characterization to biodistribution, clearance, and observed effects. The dot's brightness, once the headline feature, becomes only one line in a table of requirements.

Anecdotes from the early commercialization of display nanocrystals capture the mood of this regulatory engineering. When cadmium-based materials first began appearing in high-end screens, many engineers viewed the compliance issue as a packaging problem: keep the nanocrystals sealed, prove that cadmium is not released, and the benefit will justify the paperwork. That view was not irrational; it echoed the logic of many hazardous materials used safely inside devices. But the existence of time-limited exemptions changed the psychology. It was no longer enough to demonstrate safe encapsulation for today's product. Companies had to plan for what happens if the exemption expires. Investment began flowing into cadmium-free alternatives not only because of consumer anxiety, but because a legal sunset turns technical debt into an urgent engineering task.

The most revealing aspect of this story is that regulation does not simply choose winners by banning losers. It creates a multi-dimensional selection pressure. Materials that are marginally worse optically can win if they are easier to characterize, easier to manufacture reproducibly, and easier to defend legally. Device architectures that reduce potential exposure can become more attractive because they reduce the need for exemptions or simplify documentation. Conversely, a material that is spectacular on a laboratory bench can stall if it is hard to specify precisely, hard to keep consistent, or hard to fit into existing legal categories.

For the public, these forces are mostly invisible. A consumer comparing screens sees colour, brightness, price, and brand. Behind that purchase is an argument inside companies about thresholds, exemptions, reporting, test methods, supplier audits, and whether a particular nanocrystal is worth the compliance complexity. In that argument, regulation behaves like a silent co-designer, pushing on the same levers as physics and chemistry: composition, coatings, stability, and how tightly a process can be controlled.

A useful way to think about it is to imagine a design meeting where nobody mentions the word "law," yet every choice assumes it. "Can we source this at scale?" often really means "Can we source it with traceability and documentation?" "Is this stable?" often means "Will it stay within the defined nanoform and performance window long enough to satisfy product claims and safety assumptions?" "Can we ship globally?" often means "Can we survive the strictest jurisdiction without redesigning the product?"

Regulation, at its best, does not freeze technology; it forces clarity. It demands that companies specify what they have made, show that they can make it again, and take responsibility for the ways materials move through workplaces, bodies, and waste streams. That demand can feel like friction to inventors. It can also be the pressure that turns a clever demonstration into a durable technology that does not rely on trust alone.

10.3 Heavy-Metal-Free Dots: The Search for Bright Without Bitter

Take away cadmium or lead and the problem should vanish, right? For anyone who has ever swapped a broken part in a machine, that instinct feels natural: remove the troublesome component and replace it with a safer one that does the same job. Nanocrystals resist that kind of substitution. Their performance is not a single dial labelled "brightness" or "purity" that can be turned up with better ingredients. It is a precarious agreement between a core, its protective layers, and the surrounding device. Change the ingredients and the agreement has to be renegotiated from scratch.

The desire for heavy-metal-free emitters is easy to sympathize with. Cadmium and lead have reputations earned over centuries: as pigments that poisoned painters, as plumbing that harmed children, as industrial contaminants that left long shadows over communities. Once nanocrystals escaped the laboratory and moved toward mass-market screens and lighting, it was inevitable that regulators, brands, and consumers would ask for versions without the headline hazards.

Yet there is a second, less intuitive truth running alongside that desire: a dot without cadmium is not automatically a dot without risk, and it is not automatically a dot that can be manufactured, shipped, and kept stable inside a product for years. "Safer elements" do not guarantee safer objects. They only change the starting point.

Among the alternatives, indium phosphide (InP) has become a major cadmium-free platform for visible and near-infrared photonics and for QLEDs. It holds a tantalizing promise: a semiconductor system that can be tuned across much of the visible spectrum without relying on cadmium-based cores. For display makers under pressure to avoid restricted substances, InP looks like the obvious candidate.

In practice, achieving cadmium-like performance with InP has required deep control of defects, shells, and the chemistry at the particle's boundary with its environment. This is not a minor technicality; it is the central difficulty. The optical virtues that made cadmium selenide such a workhorse—narrow emission, high efficiency, relatively forgiving synthesis—were not gifts of quantum confinement alone. They were gifts of a materials system whose surfaces could be tamed with known tricks. With InP, the tricks have been harder to perfect.

A useful way to frame the challenge is to remember what light emission demands. When a dot absorbs energy, it creates an excited electron and a corresponding "hole" (an absence of an electron) that must find each other again. The best outcome is a radiative recombination: the pair reunites and emits a photon. The worst outcome is a detour into defects: the energy turns into heat without light. Defects are not exotic in nanomaterials. A particle only a few nanometres across has a large fraction of its atoms on or near the surface, where bonds are incomplete and chemical reactions are easier. The surface therefore becomes a field of traps unless it is carefully passivated.

Passivation means saturating or protecting those reactive, trap-prone sites so that the excited carriers are not snagged before they emit light. In core—shell designs, a shell can also serve as a physical and electronic barrier, helping keep the excited carriers away from surface imperfections. The same strategies that help toxicity by containing a hazardous core also help performance by insulating the emissive region from the messy chemistry outside. But those strategies are not equally easy for every materials system.

InP exemplifies this. Building an InP core with the right size distribution is already a more delicate chemical act than many cadmium-based syntheses. Growing a shell that fits well—meaning a shell that does not introduce strain, dislocations, or additional defect states—can be even more demanding. A shell that is perfect on paper can be imperfect in a flask, and nanometre-scale imperfections are enough to spoil the very feature you care about: emission efficiency and colour purity.

This is why InP QLED development faces both material and device-architecture challenges. The dot itself must be improved through better core/shell growth and better ligands, but the device stack also has to be redesigned so that charges arrive in a controlled way and do not damage the dot. In other words, replacing cadmium-based nanocrystals with InP-based ones has often required not just a new material, but a new relationship between the material and its electronic environment. A dot that is slightly more fragile to electrical stress forces changes in the layers that inject electrons and holes. A dot that is more sensitive to oxygen forces changes in encapsulation. A dot that has a broader emission linewidth forces changes elsewhere in the display system to hit colour targets.

That interdependence helps explain a pattern that outsiders sometimes mistake for procrastination. Why not switch immediately if the motivation is safety? Because for a company selling millions of displays, a switch that reduces performance by a small percentage can be commercially fatal, and a switch that reduces lifetime is a customer-service disaster. A screen that looks slightly duller on a showroom wall loses to a rival. A screen that shifts colour after a year becomes a warranty problem. Safety-driven substitution has to compete in the same marketplace as every other design choice.

A concrete example makes the trade-off feel less abstract. Consider the modern television showroom, where a premium display is judged in seconds. The most striking scenes are saturated reds, bright greens, and high contrast in mixed lighting. Cadmium-based dots earned their place partly because they could produce narrow, vivid colours with high efficiency, and they could keep doing it under the intense blue illumination used in backlights. InP-based alternatives have improved dramatically, but the road has involved years of work on shells and ligands to reduce the defect-related losses that blur colours and waste energy. Each incremental fix is a chemical negotiation: adjust the shell growth conditions, and you may improve stability but broaden the emission; change the ligand, and you may improve compatibility with a polymer film but worsen electrical performance in an LED device.

There is also a quieter pressure that comes from manufacturing. A laboratory synthesis can tolerate heroics: a skilled chemist watching for subtle changes in viscosity or colour, tweaking temperatures by intuition, accepting lower yields to get a clean sample. Industrial production cannot. It needs recipes that survive scaling, supply variations, and operator changes. For substitution to succeed, the heavy-metal-free material has to be not only good but *repeatably* good. A batch-to-batch shift that a research group might shrug off as normal variability becomes a serious problem when the downstream product is calibrated to precise colour coordinates.

This is where the phrase "better materials" can mislead. A material is not better in the abstract; it is better for a particular manufacturing pathway and device architecture. An InP dot that is stable in a bottle is not necessarily stable under the light and heat inside a display. A dot that looks excellent in a film may not survive the electrical stresses in an emissive QLED. The easier it is to make a dot in one context, the more surprising its failures can be in another.

The "heavy-metal-free" label can also obscure new kinds of uncertainty. Removing cadmium reduces a familiar hazard, but it does not guarantee benign biological behaviour. In vivo results show that biodistribution and persistence can be significant and depend on surface functionalization. Put more plainly: in an animal, InP-based particles with different surface groups can travel differently, lodge in different organs, and remain for different lengths of time. That is not an argument against using them; it is a warning against treating the periodic table as a simple moral map.

This point is easy to miss because public conversations often treat toxicity as a property of elements rather than objects. Cadmium has a bad name; indium does not. But a nanoscale object is an engineered system, and biology responds to systems. A well-coated cadmium-based particle might release little cadmium under certain conditions, while a poorly designed cadmium-free particle might persist in an organ and cause stress through other mechanisms. The relevant comparison is not "element versus element" but "complete design versus complete design," evaluated over time in realistic environments.

The history of substitution in other technologies offers a sobering parallel. Replacing one problematic ingredient often shifts the burden elsewhere: to a different environmental concern, to a different supply risk, or to a different performance compromise. Indium itself is not a trivial resource. It is used widely in electronics, and its supply is tied to mining and refining streams with their own environmental footprints. A global push toward indium-based nanocrystals could bring new attention to sourcing and recycling, not because indium is the next cadmium, but because large-scale materials adoption always reveals hidden dependencies.

The engineering response has been to treat safety and performance as coupled design goals rather than competing ones. For cadmium-containing dots, toxicity mitigation strategies often center on improving physicochemical design—better shells, tighter ligand binding, improved stability—highlighting that engineering controls can shift risk even when composition is unchanged. The same mindset applies to cadmium-free dots. The objective is not merely to remove a restricted element, but to produce a particle that stays intact under the stresses of manufacturing and use, that is compatible with the polymers and solvents of device fabrication, and that can be characterized and reproduced reliably.

That coupling has an emotional dimension as well. In consumer technology, trust is fragile and often binary: a material is either "safe" or "dangerous" in popular imagination, with little appetite for nuance.

Engineers and chemists, meanwhile, live inside nuance. They know that a particle's behaviour depends on its coatings, on its environment, on how it ages. When a company decides to adopt a heavy-metal-free alternative, part of what it is buying is not only a different chemistry but a different story it can tell the public and regulators. That story has value, and it can be deserved, but it still has to be backed by real stability and real lifecycle thinking. A reassuring label cannot substitute for data.

Some of the most interesting scientific creativity in the field has come from this pressure. InP chemistry has driven new approaches to shell growth and ligand design precisely because the older, simpler recipes did not deliver what screens and LEDs demand. Device engineers have had to rethink charge transport layers and architectures to accommodate different dot behaviours. The result is a broader, more diverse toolkit for making nanoscale emitters work in the messy conditions of real products.

The temptation is to treat this as a linear moral tale: first there were hazardous bright materials, then safer bright materials replaced them. The reality is closer to a continuing negotiation among competing goods. A display that saves energy by being more efficient has an environmental benefit. A material that avoids a restricted substance reduces a particular hazard. A device that lasts longer reduces waste. These benefits can align, but they can also pull against each other in the short term, and the engineering challenge is to make them align without pretending the trade-offs never existed.

The most honest promise of heavy-metal-free dots is not that they erase risk, but that they allow society to make different choices about which risks it is willing to carry. They shift attention toward defects and stability, toward how surfaces are controlled, toward how devices are encapsulated, toward sourcing and end-of-life. They also remind us, in a very practical way, that progress in materials science is rarely a clean swap. It is a careful rebuild, atom by atom, of the conditions under which brightness can be trusted.

10.4 Responsible Innovation: Trust Is a Materials Property, Too

Trust is usually described as a feeling: confidence in a brand, faith in institutions, a sense that experts are telling the truth. With technologies that are invisible to ordinary senses and difficult to evaluate without specialised tools, trust becomes something more concrete. It begins to resemble a property that can be built into a material system, supported by measurement, documentation, and habits of disclosure. A glowing nanocrystal in a screen or a vial is not only a triumph of chemistry; it is a compact between the people who make it and the people who live with its consequences.

The compact is easiest to see when it fails. The last half-century contains enough cautionary tales to make any new material feel guilty until proven innocent: industrial chemicals that seemed harmless until long-term exposure was studied, medical devices recalled after unexpected interactions with the body, pollutants that travelled far beyond the communities that produced them. None of these stories maps perfectly onto nanocrystals, but they shape the public atmosphere in which nanocrystals must operate. When people hear "nano," they do not only imagine the magic of miniaturisation; they also imagine unseen particles crossing boundaries that used to feel secure—lungs, skin, placentas, watersheds.

In that atmosphere, responsible innovation is not a moral bonus. It is the difference between a technology that is permitted to mature and one that is permanently on trial. The National Academies framing of responsible development captures the essence: balancing benefits and harms requires proactive environmental, health, and safety research, standards, and public engagement. Trust depends on anticipation and mitigation. The key word is *proactive*. Waiting for harm to appear and then reacting may satisfy a minimum legal duty, but it rarely satisfies a social one.

For nanocrystals, anticipation has a particular meaning. The danger is not just in what the material does on day one, under perfect conditions, in the hands of careful scientists. The danger sits in the gaps: between a pristine sample and a weathered one, between a controlled factory and an informal recycling stream, between a small clinical trial and widespread use. Responsible innovation tries to close those gaps early enough that society does not have to learn by being harmed.

The OECD's Safe(r) Innovation approach makes this practical rather than preachy. It positions safety integration early and across the value chain as part of sustainable innovation and treats uncertainty reduction as a governance goal. That phrasing matters: uncertainty is not merely an annoyance for regulators; it is a form of risk in itself. When the public cannot tell what a material is, where it goes, or how it changes, suspicion fills the space where information should be. Reducing uncertainty—through better characterization, clearer disclosure, and realistic exposure scenarios—becomes a way of engineering trust.

A good place to begin is with something that sounds dull: measurement. In everyday life, measurement is a private convenience. In technology governance, it is a public language. Trustworthy nanomaterial evidence depends on standardized, reproducible characterization and testing, because without it there is no stable meaning to words such as "particle size," "coating," or even "same material." OECD TG 125 and ISO/TR 13014 are examples of this governance infrastructure. They function like grammar books for nanoscale evidence: they help ensure that when one laboratory reports a size distribution or a dissolution behaviour, another laboratory can understand what was done and reproduce it.

That might seem remote from ethics, but it is the opposite. When a company claims that a particular dot formulation is stable in a polymer film, or that a biomedical formulation clears from the body in a certain way, the claim rests on measurements. If the measurements are not comparable, the claim cannot be evaluated, and the public is forced into a crude choice: trust the manufacturer or distrust them. Standardization offers a third option: trust the process by which claims are checked.

The need for this "trust stack" becomes clearer once you notice how much of the academic literature is hard to use for real risk assessment. A risk-assessment-oriented view of the field has long pointed out that usability is limited when characterization is incomplete or exposure scenarios are unrealistic. It is not that academic studies are dishonest. It is that academic incentives often favour novelty over completeness. A paper may demonstrate an exciting biological effect without fully specifying the particle's coating integrity, aggregation state, or how the test medium changed the surface over time. Regulators, in contrast, need structured data packages. They must be able to link a material identity to a dose, a route of exposure, and an outcome, and they must be able to compare results across studies.

This difference in incentives is one reason responsible innovation cannot be outsourced entirely to universities or to government agencies. Firms that commercialize nanocrystals sit at the point where the evidence has to become decision-ready. They also control many of the variables that matter most: exact formulation, process conditions, and real-world use scenarios. If they do not invest in high-quality characterization and realistic testing, the knowledge gap persists, and the public ends up as an unwilling test environment.

Medicine makes the trust stack visible because it is enforced. FDA guidance for nanomaterial-containing drugs emphasizes product-specific assessment and manufacturing controls. The language can sound procedural—quality systems, characterization requirements, process validation—but the ethical content is straightforward: if a material is going into bodies, the maker must show not only that it works, but that it can be made consistently and that its critical properties are understood. Here, trust is not granted by reputation; it is demanded by a system designed to prevent wishful thinking from becoming clinical reality.

A concrete example helps. Imagine a nanocrystal-based imaging agent proposed for tumour visualization. The scientific excitement is easy to understand: brighter labels, longer observation times, multiplexed colours. The regulatory questions arrive immediately: What exactly is the particle's size distribution? How stable is the coating in blood? Does the formulation contain free ions? What impurities remain from synthesis? Where does the agent accumulate, and how long does it persist? Can the manufacturer produce the same formulation batch after batch, with the same behaviour in vivo? Those questions are not rhetorical hurdles. They are a way of translating a laboratory marvel into something that can be ethically offered to patients, who cannot personally audit a nanoparticle's chemistry.

The same logic applies to consumer products, though the enforcement mechanisms differ. People do not consent to a display in the same way they consent to a drug, and they do not receive individualized benefit from a screen in the way a patient might from a diagnostic agent. That changes the ethical balance. For consumer electronics, the justification is usually collective and diffuse: better energy efficiency, better colour, thinner devices, new capabilities. The burden, meanwhile, can fall on particular workers, communities near manufacturing sites, and communities managing waste. Responsible innovation has to make these distributions visible.

Worker protection is the first, most immediate justice issue. A consumer may never contact free nanocrystals in a well-encapsulated device. A worker handling dispersions, cleaning reactors, maintaining filters, or dealing with waste streams might. Ethical design therefore includes mundane choices: whether to keep materials in liquid form rather than powders; whether processes are enclosed; what ventilation and protective equipment are standard; how spills are handled; how waste is labeled and treated. These are not merely compliance details. They are ways of deciding whose lungs and skin will bear uncertainty.

The second justice issue is end-of-life. As earlier discussion implied, containment during use does not guarantee containment in disposal. Responsible innovation asks what happens when a device is shredded, burned, dumped, or recycled. It also asks where those processes occur. Formal recycling systems with controlled conditions are one world. Informal recycling, where devices are dismantled by hand and burned to recover metals, is another. If a technology's benefits accrue largely to affluent consumers while disposal burdens are exported to poorer communities, the moral ledger does not balance, no matter how impressive the colours look.

This is where lifecycle accountability becomes more than a technical phrase. Accountability means designing products so that materials can be recovered or at least managed safely, and it means contributing to systems that can actually do that management. It also means resisting the temptation to declare safety simply because a material is "encapsulated." Encapsulation is valuable, as already discussed, but it can be a temporary condition. A responsible claim respects time, heat, light, and human behaviour.

Transparency is the third pillar, and it is often the most contested because it touches corporate secrecy. People understandably want to know whether a product contains restricted elements or novel nanomaterials. Companies understandably worry that detailed disclosure reveals proprietary formulations. Between these interests sits a practical compromise that responsible innovation pushes toward: clear disclosure of material classes and hazards, along with accessible summaries of independent testing and lifecycle management. The goal is not to publish trade secrets; it is to avoid forcing the public into blind trust.

Independent testing matters because it changes the social meaning of a claim. When a manufacturer says “safe,” the public hears advertising. When an independent lab reports standardized measurements using recognized methods, the public hears evidence. The difference is not only psychological; it changes accountability. Independent testing creates a record that can be debated, improved, and corrected.

Communication about uncertainty is part of transparency, and it is where honesty is hardest. The public often wants certainty: safe or unsafe, yes or no. Scientists rarely have that luxury, especially with long-term environmental fate and low-dose chronic exposures. Responsible innovation does not promise impossible certainty. It promises seriousness: a willingness to say what is known, what is not known, what has been tested under realistic conditions, and what is being done to reduce the unknowns.

Regulatory regimes such as TSCA and REACH operationalize responsibility in precisely this way: through reporting, identity and characterization requirements, and safety documentation that travels with the material. These regimes can feel burdensome, and they can be, but they also create a shared baseline. They reduce the ability of irresponsible actors to hide behind ambiguity and they reward organizations that invest in evidence.

There is a historical lesson here from the early years of nanotechnology policy. In the 2000s, "nano" became a catchword that attracted funding and excitement. It also attracted a wave of public concern, partly because the term was broad enough to include almost anything small, and partly because memories of earlier chemical controversies were fresh. One response was to insist that nanomaterials were simply materials, nothing new, and that existing safety frameworks were sufficient. Another response was to demand sweeping bans. Neither extreme helped innovation mature responsibly. What helped was a slow build-out of characterization standards, reporting expectations, and more careful thinking about exposure routes and lifecycle transformations. That work is not glamorous, but it is what makes it possible for a technology to be evaluated rather than merely feared.

Fairness in access adds another dimension to trust. If nanocrystal-based imaging improves diagnosis, who receives it first? If it reduces energy consumption in lighting or displays, who benefits from lower bills and better performance? If manufacturing shifts to regions with weaker protections, who bears the hidden costs? These questions do not have purely technical answers, but technical choices influence them. A formulation that requires rare precursors and tight manufacturing conditions may raise costs and limit access. A design that enables recycling can reduce long-term burdens on communities handling waste. A product that can be serviced rather than discarded changes the waste stream. Ethics, here, is not separate from design; it is expressed through design constraints and priorities.

A final aspect of responsible innovation is humility about substitution. Switching to cadmium-free materials can be a meaningful improvement, but it does not eliminate the need for careful evidence. "Heavy-metal-free" does not mean "biologically inert," and it does not guarantee benign environmental fate. Trust is damaged when labels are used as absolutes. Trust grows when labels are treated as starting points for ongoing evaluation.

The practical outcome of all this is that responsible innovation looks like a set of disciplines rather than a slogan: characterize materials with the rigor needed for comparability; test under exposure scenarios that resemble real use and real disposal; build manufacturing control so the product is what it claims to be; protect workers as the first line of ethical responsibility; design for end-of-life rather than hoping someone else will manage it; disclose enough for the public and regulators to judge claims; and treat uncertainty as something to reduce, not something to hide.

A nanocrystal's glow is an achievement of precision. Responsible innovation asks for a matching precision in how society manages the costs and benefits of that glow. When that precision is present—in measurement, in transparency, in protections for the people closest to exposure, and in honest accounting for a product's afterlife—trust stops being a public-relations wish and becomes a property that has been engineered into the technology itself.

Chapter 11

From Nanocrystal to Infrastructure: Factories, Standards, and the Next Quantum Leap

A nanocrystal's color can be tuned with exquisite precision, yet that elegance means little until millions or billions of such particles behave predictably outside the lab. Here the drama shifts from discovery to dependability: recipes become manufacturing protocols, surfaces become engineering bottlenecks, and the same size-based quantum rules that transformed screens begin to point toward devices that emit not many photons, but exactly one.

11.1 Scaling the Recipe: When Chemistry Becomes Process Engineering

A laboratory synthesis can be thrilling in the way cooking for friends can be thrilling: the heat is high, the timing is tight, the glassware fogs and clears, and at the end there is a vial that catches the light and answers with a clean, confident color. On the bench, the chemist's hands and habits are part of the apparatus. At scale, those hands have to disappear. A factory cannot depend on whether someone swirled the flask with just the right wrist motion, or whether the oil bath happened to recover its temperature quickly after a cold syringe went in. The romance gives way to something more austere and, in its own way, more impressive: the discipline of making the same thing again and again, on bad days as well as good ones, across shifts, across sites, across months.

That shift is sharper for nanocrystals than for many other materials. A quantum dot is small enough that a handful of atoms more or less is not a rounding error; it is a design change. And because so much of a nanocrystal is surface, the dot is never just its inorganic core. The *skin*—the ligands, the charged groups, the dangling bonds, the quiet contamination picked up during purification—is not decoration. It is chemistry that controls how the dot grows, how it survives, and how it behaves once it is asked to do work inside a device.

The temptation, especially for outsiders, is to treat color as the product specification: green at 530 nm, red at 630 nm, with a narrow peak. Factories do care about that, of course. But as soon as a dot becomes a component rather than a demonstration, "same color" is only the beginning of "same material." Two vials can share an almost identical photoluminescence spectrum and still diverge in ways that only show up later: one batch bright in a film but fragile under heat; another stable in air but reluctant to pass charge; one with a high initial quantum yield but a slow, steady decline over weeks. The spectrum is a signature, but it is not a full biography.

The core problem hiding beneath the language of recipes is that nanocrystal synthesis is a contest between two processes that have to be timed just right: nucleation and growth. In the canonical hot-injection style, the goal is to create a sudden burst of nuclei—tiny seeds—and then feed those seeds in a controlled way so that they grow without spawning a second generation of newcomers. That "burst then grow" idea is elegant on paper and workable on a bench. When the volumes get larger, however, the physical act of mixing becomes a deciding variable. Heat has inertia. Fluids form gradients. A large reactor is not a scaled-up vial; it is a different physical environment.

One useful way to see the difference is to imagine the moment of injection, when a precursor meets a hot solvent. In a small flask, the cold stream warms quickly, and the whole volume recovers. In a large vessel, the cold stream can create a transient local pocket: cooler, more concentrated, with different chemistry for a few seconds. Those seconds can be enough to produce a tail of too-small particles, or to encourage an unwanted secondary nucleation event. The average temperature might be "correct," but the dots do not live in the average. They live in local conditions, and they remember them.

Industrial thinking replaces the single story of "the reaction" with a distribution of stories happening simultaneously across space and time. The target is no longer merely a mean particle diameter; it is a tight spread, a predictable tail, and a low rate of rare outliers that can later become failure seeds. This is where chemistry becomes process engineering. The factory asks questions that sound less romantic but are more honest about causality: What is the residence-time distribution? How fast does the reactor mix relative to the kinetics of nucleation? Where are the thermal gradients, and how stable are they from run to run? If a pump drifts by 2%, what does that do to the dot?

For quantum dots, the translation from bench to plant often proceeds through the idea of continuous flow, which sounds like a mere change in plumbing but is really a change in philosophy. Batch synthesis is episodic: one moment of injection, one period of growth, one quench. Continuous reactors treat production as a controlled stream. A flow rate becomes a dial. A reactor's geometry becomes a hidden reagent.

Heat exchange becomes a primary lever rather than background. In some designs, the hottest zone is reserved for nucleation, and then the mixture is quickly moved to a cooler zone optimized for growth. The point is not that continuous flow is automatically superior; it is that it makes time and temperature trajectories easier to specify and repeat, and it can reduce the run-to-run personality of a reaction.

This is also where the word "scale-up" can be misleading. The dream is to keep everything the same and make more of it. Nanocrystals rarely allow that. When the physical environment changes, the control knobs change. Stirring speed, injection rate, reactor diameter, heat-transfer area, and even how quickly a system can be cleaned between runs can shape what "the same chemistry" means. A plant that makes reliable dots is not simply repeating a chemical recipe; it is stabilizing a coupled system of chemistry, transport, and measurement.

Measurement deserves to be named as part of the manufacturing process, not merely its audit. A research spectrometer can tell you where an emission peak sits and how sharp it is. Manufacturing needs more: a fingerprint of the particle population and its surface state. That might include absorption features, time-resolved photoluminescence (how long excitations live before emitting), and indicators of trap-assisted recombination, where defects steal energy that should have become light. It might include chemical assays for residual precursors, free ligands, or ionic contaminants that can later catalyze degradation. It often includes microscopy and scattering measurements to catch aggregation—not just whether it happens, but whether it is beginning to happen.

None of these measurements is glamorous, and that is why they are so revealing. They answer questions a spectrum cannot. A batch can be tuned to the right peak wavelength by nudging growth time or temperature, yet still carry a different internal distribution of defects. The defect landscape is a phrase that sounds abstract until one imagines what a defect does. In a semiconductor nanocrystal, a defect can behave like an emergency exit for energy: an electron and a hole (the pair created by absorbed light) meet, but instead of recombining radiatively and producing a photon, they lose their energy as heat through a non-radiative pathway. In display and lighting applications, that is lost brightness and lost efficiency. In imaging, it is reduced signal. In the long run, defect sites can also become chemical footholds where oxygen or moisture begins a slow attack.

The surface is where many of these exits are located. A dot's ligands passivate surface atoms, satisfying bonds that would otherwise be "dangling" and electronically active. But ligands are not a single thing. Their head groups bind with different strengths; their tails pack differently; they can exchange during purification or storage; they can reorganize in a film. Small changes in ligand shell structure can shift growth kinetics during synthesis and also shift compatibility with later processing. In other words, the factory has to control not only the nanocrystal core, but the *social life* of its surface molecules.

A concrete example of how this plays out can be found in the display industry's long courtship with quantum-dot enhancement films for LCD backlights. These films depend on dots dispersed in a polymer matrix and excited by blue LEDs. On paper the requirement is simple: convert blue to narrow red and green with high efficiency. In practice, manufacturers learned that the same nominal red dot can behave differently when embedded in a film and driven hard for thousands of hours. One batch may have a perfectly acceptable initial spectrum and brightness but show faster degradation in the presence of trace oxygen permeating through a barrier layer. Another may resist oxidation better but be more prone to aggregation, producing haze or scattering that reduces optical performance. The consumer sees a shift in white balance or a loss of brightness; the factory sees yield loss, warranty risk, and the uncomfortable truth that "identical" batches were not identical in the ways that mattered.

Once a material enters products, time becomes a manufacturing parameter. A dot that survives a day on the bench has not yet proved it can survive a year in a device, or five years in a television that sits in sunlight, warms and cools daily, and endures occasional power surges. Long-term tests are slow and expensive, so industry searches for accelerated stress tests: heat, humidity, intense illumination, electrical bias. But accelerated tests can mislead if they trigger degradation pathways that are not dominant in real use. Process engineering here becomes statistical and epistemic: not only how to make a batch, but how to know, quickly, whether it will behave like the batches that have already earned trust.

This is where statistical process control enters the story, a set of ideas born in older industries and surprisingly well suited to nanomaterials. Instead of treating each batch as an isolated success or failure, the factory treats production as a stream of data. The question becomes: are key parameters drifting? Are they stable within a known window? Are rare excursions appearing more frequently? Control charts and process capability metrics are not fashionable words, but they have a moral dimension in high-volume manufacturing: they replace hope with quantified vigilance.

To apply them, however, one must decide what to measure. A convenient metric that correlates poorly with performance can become a trap. Peak wavelength is easy to record and easy to optimize. Quantum yield is more meaningful but can be measured differently depending on geometry, reabsorption, and calibration. A time-resolved decay curve can reveal subtle increases in non-radiative recombination that a simple brightness value hides. Particle size can be inferred from optical features, but optical similarity can mask changes in shape or surface disorder. In manufacturing, metrology has to be designed with the end use in mind. A dot intended for color conversion in a film is judged by photostability and behavior in a matrix; a dot intended for electroluminescence is judged by charge balance, interface compatibility, and resistance to electrically induced degradation. "Good" is not a universal label; it is a contract with an application.

Purification and separation, too, turn out to be part of the contract. Bench chemistry often treats purification as cleanup: remove unreacted precursors, excess ligands, and side products. In a factory, purification is where a significant fraction of cost, yield, and consistency is determined. Methods that are acceptable at small scale can become punishing at large scale. Each washing step risks losing material, exchanging ligands, or introducing trace impurities. Yet insufficient purification can leave behind species that quietly poison stability.

Separation is not merely about cleanliness; it can be about narrowing the distribution of particles produced by the synthesis. In practice, some of the most uniform nanocrystal populations are not born uniform; they are *made* uniform by sorting. Size-selective precipitation, filtration, and chromatography can all be used to trim tails and remove aggregates. These methods carry a manufacturing trade-off: the tighter the specification, the more material is rejected or recycled, and the more complex the process becomes. The factory has to decide how much variability to tolerate in synthesis and how much to correct downstream. That decision is economic as much as scientific.

A historical anecdote from chemistry helps here, because this pattern is not unique to quantum dots. The synthetic dye industry in the nineteenth century learned, sometimes painfully, that a new molecule's color in a flask did not automatically translate into a reliable textile dye. The same nominal dye could vary with impurities that changed fastness—how well the color held under washing and sunlight. Manufacturers discovered that purification, standardization, and testing were not peripheral; they were the difference between a curiosity and an industry. Quantum dots repeat this lesson at the nanoscale. Color is the headline, but stability and reproducibility are the business.

The tyranny of small variations is amplified by the way dots are used. A modern display is not a single emitter; it is millions of pixels or a large-area film with microscopic uniformity demands. The human eye is remarkably sensitive to small differences in color and brightness across a screen, and cameras and automated inspection systems can be even more unforgiving. A batch that is "mostly fine" can still be disastrous if it contains a small fraction of particles that aggregate or quench, creating specks, streaks, or long-term drift. The language of manufacturing is therefore often the language of tails: not the average dot, but the rare dot that causes a defect, and the rare defect that causes a return.

At this point the word "defect" starts to mean two things at once. There are defects in the nanocrystal, which affect optical and electronic behavior. And there are defects in the manufacturing process, which create non-uniformity in films, contamination, or misprocessing. Process engineering tries to suppress both by imposing repeatable conditions and by making every step legible. If a dot fails in a device, the factory wants traceability: which lot of precursors, which reactor conditions, which purification train, which operator shift. The aim is not surveillance for its own sake. It is the ability to learn.

One of the more surprising psychological changes in this transition is the redefinition of "skill." In a research lab, a skilled chemist can coax a synthesis into working, adjusting by intuition when a solvent behaves oddly or a precursor seems sluggish. In a factory, that kind of improvisation is dangerous. Skill becomes the ability to design a process that does not require heroics. A good manufacturing process is one in which an ordinary day produces extraordinary consistency.

This is also why manufacturing pushes researchers toward thinking about mechanisms that are boring but decisive: heat transfer coefficients, mixing regimes, pump pulsation, fouling of reactor walls, the kinetics of ligand exchange. A reactor wall that gradually accumulates deposits can change the effective surface chemistry of the system, creating nucleation sites or scavenging ligands. A slight change in precursor purity can alter nucleation timing. A ligand batch from a supplier can vary in chain-length distribution and therefore in how it packs on the surface. These are not mysteries; they are mundane realities that become visible only when one tries to make the same thing a thousand times.

There is a temptation to frame all this as industry's cold intrusion into science's playful creativity. A better view is that scale reveals a deeper layer of the science. If a small change in mixing creates a measurable change in defect populations, then mixing is not just engineering; it is part of the chemical environment that determines which atomic arrangements are favored. If a ligand's packing changes growth kinetics, then the ligand is not merely a stabilizer; it is part of the thermodynamic and kinetic landscape that shapes the nanocrystal. Reproducibility is not merely an industrial obsession. It is a microscope aimed at causality.

In the end, "chemistry becomes process engineering" because the object being manufactured is not just a particle but a performance. The factory is asked to deliver not merely a wavelength but a set of coupled behaviors: brightness, stability, compatibility with encapsulants and polymers, resistance to heat and light, predictable interactions at interfaces. Achieving that means treating variation as the central adversary. The work is not only to hit a target, but to narrow the spread, clip the tails, and keep the target from drifting over time.

That discipline changes how one thinks about invention. A new nanocrystal composition can be genuinely better in the lab and still be a poor industrial candidate if it demands impossible control over moisture, or if its synthesis has an uncomfortably narrow window, or if its purification requires steps that are too costly or too wasteful. Conversely, a material that is slightly less spectacular on a single metric can win if it is forgiving, consistent, and robust under real processing. Industrial success is often the victory of the second-best idea that can be made reliably.

A vial glowing on a bench is a promise. A factory that makes that glow dependable turns the promise into infrastructure: a component whose behavior is stable enough that other people can design around it without fear. When that happens, the dots cease to be a laboratory marvel and become something quieter and more consequential: a material with a specification, a supply chain, and a track record. The glow remains, but now it carries the weight of a process that can be trusted.

11.2 Printing Light: Integration as the Final Experiment

A vial of nanocrystals in a glovebox is a controlled world: oxygen kept out, moisture kept out, light levels chosen, temperature steady. A finished display, by contrast, lives in a bright, warm, occasionally humid universe full of adhesives, polymers, metal oxides, and mechanical stresses. The hard truth of commercialization is that a material can be a triumph of synthesis and still fail at the moment it is asked to become a layer in someone's architecture. Integration is unforgiving precisely because it is ordinary. It forces a luminous nanocrystal to share space with the rest of manufacturing.

The first shock is that the material is no longer "a particle." It becomes an ink. An ink is a negotiated truce among competing demands: it must flow through a nozzle or spread under a blade; it must wet one surface but not creep onto another; it must dry fast enough for throughput but slowly enough to level; it must leave behind a uniform film without pinholes, ridges, or coffee-ring staining. It must do all that while keeping the nanocrystals bright and separated from one another, and while keeping their ligand shells intact enough to prevent new non-radiative pathways from opening.

A second shock is that the most important chemical reactions may be the ones nobody planned. A ligand that behaved politely in solution can exchange with molecules in a host polymer. A residual solvent can plasticize a layer that was supposed to be rigid. An additive that improves printability can introduce impurities that later catalyze degradation. Interfaces become a kind of chemical borderland: a place where small differences in surface energy, charge, and local composition can decide whether a film works or quietly deteriorates.

The central use-case that has turned integration into a practical art is display manufacturing. Quantum-dot-based color conversion is a leading strategy for full-color micro/mini-LED displays. The idea is straightforward at the level of a cartoon: start with efficient blue μLEDs, then convert some of that blue light to green and red in patterned regions, so that each pixel produces the full-color triad. The conversion layer must be placed with micrometre-scale precision, at high yield, across large areas, and it has to keep performing after long hours of intense blue excitation.

The moment those requirements are stated, the material stops being a curiosity and starts behaving like infrastructure. Every step—ink formulation, deposition, drying, patterning, encapsulation—becomes part of the "material," because every step can create or eliminate defects that the eye will notice.

A layered device stack makes clear why integration is the final experiment: the nanocrystals sit inside a sandwich of other layers, each with its own chemistry, and each interface is a place where brightness can be gained or lost. The protective barriers on top are not mere packaging; they are life support. The banks that define pixels are not mere geometry; they can sculpt how liquids dry and where material accumulates. The blue μLED beneath is not simply an illumination source; it is a thermal and photochemical stress test applied thousands of times per second.

One concrete example shows how such ordinary-seeming details become decisive. Inkjet printing is practical for depositing color-conversion layers, and it has become a central workhorse for patterned deposition. Yet inkjet printing brings its own characteristic defect modes. When droplets land inside bank structures (tiny walls that keep colors separated), a printed liquid can creep up the bank edges. Engineers call this "edge climbing," and it can be visually disastrous. It produces a film that is thicker near the walls and thinner in the center, which changes conversion efficiency. Too thin, and blue light leaks through, washing out color. Too thick, and the layer may crack or scatter light, or simply waste expensive material. The point is not that inkjet is flawed; it is that it exposes how much of display performance is dictated by fluid mechanics at the micrometre scale.

Thickness targets are not aesthetic preferences; they are physical requirements. For a conversion layer to do its job, it must absorb enough blue photons and re-emit them efficiently. If the layer is too thin or too diluted, the blue light is only partially converted. If the layer is too thick, it can self-absorb its own emission, trap heat, and become more sensitive to tiny pinholes or local quenching sites. A manufacturing line therefore ends up treating thickness uniformity as a first-class specification, alongside color coordinates and brightness.

This is one reason integration can feel like a new branch of science: it forces optical physics, colloid chemistry, and printing technology into the same conversation.

Printing also changes the internal life of a film. In solution, each nanocrystal is surrounded by solvent and can rotate freely. In a solid layer, particles are close enough to exchange energy and to affect one another's emission. The host matrix—often a polymer—is not inert. It decides how well particles stay separated, how much oxygen can diffuse, and how the film responds when it heats and cools. A polymer that is too permeable to oxygen and moisture can shorten lifetime. A polymer that is too rigid can crack under thermal cycling. A polymer that interacts strongly with ligands can pull them off the surface, trading stability for new trap sites.

The end result is that "good in a vial" is a weak predictor of "good in a film," and "good in a film" is a weak predictor of "good in a finished stack." That hierarchy of tests is why integration is so often where promising materials stall.

Device makers, understandably, would prefer integration steps that fit into established toolchains. This is part of the appeal of direct photolithographic patterning of nanocrystal layers. Photolithography is the quiet engine of modern electronics, capable of high resolution and alignment across large substrates. The attraction is not only finer pixels; it is also uniformity and repeatability. But photolithographic routes are not a simple plug-in. They demand chemistry—process co-design: the nanocrystal's ligand set, the photochemistry used to crosslink or polymerize a surrounding matrix, and the solvents used in development all must be compatible. A resist process that is gentle to silicon can be harsh to colloids. Light used to pattern a layer can also photo-oxidize it. Developers and etchants can strip ligands or roughen surfaces.

So integration becomes a search for a process window: a range of conditions where patterning is sharp, films are uniform, and optical performance survives. The trade-offs are real. Printing methods such as inkjet offer simpler tooling and a natural way to place different colors in different places, but they must fight drying instabilities, nozzle variability, and interactions with bank geometry. Alternatives such as aerosol-jet printing are explored partly to expand that process window, offering different droplet formation physics and potentially different control over feature geometry, aspect ratio, and film formation. Lithographic approaches promise high resolution and alignment, but they require a level of chemical robustness that many colloids were not originally designed to handle.

In practice, this is what "integration" means: a sequence of compromises that still yields a film that passes both optical and reliability tests.

Reliability engineering enters the story most forcefully through oxygen and moisture ingress. A nanocrystal layer is vulnerable not only to catastrophic exposure but to slow diffusion through polymers, adhesives, and microscopic defects. Water and oxygen can work their way through barrier layers, especially at edges where films terminate and seams exist. A barrier that is excellent in the center of a panel can still fail through edge ingress, a mode of failure familiar across thin-film technologies. For large-area manufacturing, these are not exotic concerns; they are the mundane geometry of real products.

Thin-film encapsulation strategies are pursued to block ingress while avoiding contamination associated with some packaging approaches. The phrasing may sound bureaucratic, but it names an uncomfortable fact: packaging can introduce its own problems. Adhesives can outgas. Plastics can leach. A seal can trap solvents. If one thinks of a device as an ecosystem, encapsulation is not an external shell; it is a climate-control system that must be compatible with everything inside.

A historical analogy helps make this emotional rather than merely technical. Early color photography did not become dependable simply because chemists found light-sensitive dyes and silver halides. It became dependable because engineers learned how to build layered films in which each layer did its job without poisoning the others: developers that did not wander, couplers that stayed put, protective coatings that blocked oxygen and scratches. The leap from "it works once" to "it works for everyone" came from mastering interfaces and barriers. Modern emissive and color-conversion stacks repeat that lesson in a new medium, with new failure pathways.

The human side of this work is easy to overlook. When a display fails in the field, it rarely fails with drama. It fades. A white point drifts. A green becomes slightly yellowish. A small region dims. Consumers may blame the brand or the technology in general. Engineers blame something more specific: a microscopic population of defects in a barrier film, a solvent residue that altered polymer morphology, a ligand exchange reaction that no one intended. The failure is quiet, and that quietness makes it hard to locate.

That is why device integration relies so heavily on "hidden" measurements: tests that do not simply read out peak wavelength. Engineers interrogate films for uniformity across the panel, for haze and scattering, for pinholes, for adhesion strength, for permeability, for how emission changes under thermal load. They also care about standardized colorimetry. Claims about color gamut and emission purity depend on measurement frameworks from the CIE, which provide a shared language for what "red" and "green" mean in a device context. Without that common language, manufacturing cannot compare lines, suppliers, or generations of product.

One more subtle truth emerges when one watches a mature manufacturing line: stability is not a property that sits *inside* the nanocrystal like a gemstone's hardness. Stability is an emergent property of a whole stack. A nanocrystal can be chemically stable but still fail if the host matrix lets in oxygen. A nanocrystal can be sensitive but still survive if the encapsulation is superb and the interfaces are well engineered. A film can be bright but still disappoint if it cannot be patterned cleanly at the required resolution, or if it generates too much variability across pixels. Performance is negotiated across layers.

That negotiation also feeds back into how nanocrystals are designed. Ligands that were optimal for synthesis may be suboptimal for printing. A ligand shell that gives high photoluminescence in solution may create poor film formation, or it may electrically insulate particles in architectures that require charge transport. The integration problem thus forces a more grown-up view of "best": not the brightest material under ideal conditions, but the one that fits into an ecosystem of solvents, polymers, transport layers, and barrier films.

A blue μLED array plus patterned color conversion is, at heart, an attempt to build a machine for controlling photons with industrial precision. In that machine, the nanocrystal layer is both the star performer and the most sensitive collaborator. The astonishing part is not that failures happen; it is that the system can be made reliable at all. When it works, it works because every seemingly boring choice—how a droplet dries, how a bank edge wets, how a barrier seals at a corner, how an interface is cleaned before coating—has been tuned until the nanocrystal's brilliance survives contact with the world.

The final product hides this labor beautifully. A screen appears to be made of pure color, effortless and flat. Underneath, it is an engineered stack where chemistry, optics, and manufacturing have agreed to cooperate, at least for long enough to earn a warranty.

11.3 Single-Photon Dots: The Color Code Meets Quantum Information

A display engineer asks a nanocrystal to do something fundamentally generous: take in energy and give it back as light as efficiently as possible, millions or billions of times. Quantum information asks for something stingier and far more exacting. The dream component is an emitter that produces *exactly one* photon when triggered, not two, not zero, and not a noisy spray whose statistics merely average out in the end. Here brightness is not the main virtue. Reliability at the level of individual quanta is.

That idea can sound almost philosophical until it is tied to a mundane problem. Imagine trying to run a secure communication link in which information is encoded in single photons. If, occasionally, your "single-photon source" emits two photons, an eavesdropper can steal one and leave the other behind, and the very mechanism intended to guarantee security becomes a loophole. Or imagine building a photonic quantum computer where photons interfere like ripples on a pond. If the source sometimes emits nothing, sometimes emits two, and sometimes emits photons with slightly different colors or arrival times, the interference that carries the computation washes out. The hardware demands a new kind of honesty from materials: no averaging over huge numbers, no hiding behind "close enough," no reliance on the forgiving physiology of the human eye.

The same confinement that turned nanocrystals into a color palette also makes them plausible single-photon emitters. A photon comes from a recombination event: an electron and a hole meet and release energy as light. In a dot, those carriers are confined, so the exciton—the bound electron-hole pair—is spatially corralled into a tiny volume. That confinement increases the chance that a single absorption event leads to a single radiative recombination. It also allows the emitter to be small enough to integrate with nanoscale photonic structures, the optical equivalent of the barrier films and patterned banks that made display applications possible.

Single-photon emission is usually diagnosed by a statistical signature with an ungainly name: *antibunching*. The everyday meaning is simple. Ordinary light sources, like a lamp, emit photons in clumps and crowds. Over very short times, the probability of detecting two photons close together is not suppressed; it is typically enhanced, because emission events are uncorrelated and can happen back-to-back. A true single-photon source, however, has a built-in pause. After it emits one photon, it must reset before it can emit the next. That pause creates a telltale dip in coincidence measurements at zero delay: photons refuse to arrive in pairs. The statistics reveal an indivisible emitter.

For colloidal nanocrystals, that promise comes with immediate caveats. The first is blinking, already familiar as a nuisance in imaging: the dot turns on and off unpredictably. In a display, blinking can be averaged away because many emitters contribute to each pixel and the eye integrates over time. In single-photon operation, blinking is catastrophic. When the dot is "off," the source emits no photon at all. A source that is only intermittently a source is not a source; it is a gamble.

Blinking is closely tied to surfaces and charging. A nanocrystal can become temporarily charged by trapping an electron or a hole at a defect, often at the surface. Once charged, the dot can lose energy non-radiatively through Auger recombination, a process in which the energy that would have produced a photon is instead handed to a third carrier, like a billiard-ball collision in quantum form. Auger processes are bad news in any light emitter because they waste excitation. In single-photon devices they are worse: they add randomness. The dot might absorb energy and then, through an Auger pathway, convert it into heat rather than the one photon you needed at that moment.

This is where the story takes a turn that mirrors manufacturing. On the bench, a dot that is "bright" is a success. In quantum information, brightness must be accompanied by purity, stability, and coherence. Purity means the source really produces one photon per trigger, with a vanishing probability of multi-photon events. Stability means the dot does not blink, does not drift in brightness, and does not wander in emission energy. Coherence, in plain terms, means the emitted photon behaves like a well-defined wave that can interfere cleanly with another photon. If photons are the currency of quantum technologies, coherence is the mint mark that proves they are compatible.

Coherence sounds abstract, but it has a simple enemy: noise. Noise can come from phonons—tiny vibrations of the lattice—that blur the emitted photon's energy. It can come from spectral diffusion, in which the dot's emission energy jitters over time because charges in its environment rearrange. It can come from the ligand shell itself if it creates a fluctuating local electric field. The emitter is not a lone atom in a vacuum. It sits in a dense, restless molecular neighborhood.

This is why surface engineering becomes both more urgent and more paradoxical. For displays, ligands are chosen to keep colloids stable in solution and compatible with polymers and barrier layers. For quantum information, a ligand shell must still do those jobs, but it also has to behave as a low-noise environment. A surface that is chemically robust may still be electrically noisy. A ligand that binds strongly may still permit slow rearrangements that cause spectral jitter. The "best" surface for one application can be a poor choice for the other, not because anyone made a mistake, but because the specification has changed from "bright on average" to "predictable on demand."

The shift in specification is so stark that it creates a fork in materials strategy. Decades of work on conventional colloidal nanocrystals, such as the well-studied CdSe family, have built a deep understanding of how to tune emission, how to passivate surfaces, and how to suppress blinking. Meanwhile, perovskite nanocrystals have surged into view as especially promising for coherence-related quantum photonics phenomena. Reports of coherent emission and spin/coherence manipulation in perovskite nanocrystals have made them feel, to many researchers, like a colloidal material that might naturally speak the language of quantum optics. That promise is tempered by familiar worries: stability, sensitivity to environment, and reproducible integration. A dot can have beautiful optical properties in a controlled measurement and still be impractical as a device component if it degrades or if it cannot be placed reliably into a photonic circuit.

A historical anecdote helps convey why this matters. When the first lasers were demonstrated in 1960, they were not immediately the tidy, reliable tools now found in barcode scanners and fiber-optic networks. Early lasers were temperamental, and the deeper revolution was not merely the principle of stimulated emission but the engineering of cavities, mirrors, pumping schemes, and materials good enough to make laser light a standard commodity. Single-photon sources are often compared to lasers, but in reverse: a laser produces an avalanche of identical photons; a single-photon source produces one at a time. In both cases, the trick is not only the emitter but the infrastructure around it.

That surrounding infrastructure in quantum photonics is typically nanophotonic integration: cavities, antennas, and waveguides built to control how light is emitted and where it goes. A cavity can enhance emission into a desired mode, making the source brighter in the direction you want and faster in its response. An optical antenna can shape radiation patterns. A waveguide can capture the emitted photon and route it on-chip. Reviews of colloidal single-photon sources emphasize precisely this point: improving optical properties is necessary but not sufficient, and coupling emitters to photonic structures is a critical step toward practical devices. It is a familiar pattern by now. A dot in solution is one world; a dot embedded in an engineered environment is another.

The parallel with display engineering is more than poetic. In a display, barrier layers and encapsulation quietly determine lifetime by controlling oxygen and moisture ingress. In quantum photonics, the photonic structure and the local dielectric environment quietly determine whether a photon is usable. A source that emits into free space at random angles is hard to collect efficiently. A source that emits photons of slightly varying frequency is hard to interfere. A source that sits near fluctuating charges is hard to keep coherent. The "stack" is different, but the lesson is similar: the surrounding architecture is not an accessory; it is part of the device physics.

To see the special cruelty of the single-photon requirement, consider what "on demand" really implies. In the ideal case, you trigger the emitter with an electrical pulse or an optical pulse, and out comes one photon, promptly and predictably. If the emitter sometimes produces two photons, your single-photon purity drops. If it emits with a variable delay, timing jitter rises and synchronization becomes difficult. If the emitted photon's color drifts, interference visibility falls. These are not small degradations; they can render a whole approach unusable.

Even the concept of "brightness" changes meaning. A display engineer can pile many emitters together. A single-photon device often involves one emitter, so brightness becomes a probability per pulse: the chance that a photon appears when asked. Raising that probability often means exciting the emitter harder, which in turn can increase the likelihood of multi-exciton states. Multi-excitons are fertile ground for Auger recombination. The very act of trying to make the source more reliable can trigger the mechanism that makes it less honest.

This tension explains why single-photon research is filled with strategies that sound almost like moral education for a nanocrystal: suppress bad habits, encourage good ones, control the environment. One path is to engineer the nanocrystal so that the radiative pathway dominates and non-radiative pathways are energetically disfavored. Another is to create structures that quickly extract the photon, shortening the time window in which dephasing and spectral diffusion can occur. Another is to choose host materials and surface treatments that minimize charge noise. None of these is a single knob. They are interlocking constraints, and improving one metric can worsen another.

There is also a sobering benchmark in the field: epitaxial III–V quantum dots, grown by methods more familiar to the semiconductor industry than to colloid chemists, remain leading platforms for near-ideal single-photon performance. They can offer impressive purity and, in the best cases, photons that are close to indistinguishable—so similar that two photons from the same source can interfere as if they were truly identical. Colloidal systems, including perovskite nanocrystals, are often discussed against this backdrop. The competition is not merely academic. Epitaxial dots can be integrated with photonic structures in a deterministic way on a chip, but they are not trivial to manufacture at scale and may require expensive growth. Colloidal dots, by contrast, carry the alluring prospect of solution processing and versatile chemistry, but they must close gaps in stability, coherence, and deterministic placement.

Deterministic integration is a phrase that captures a practical headache. In a finished photonic device, the emitter has to be in the right place relative to a cavity or waveguide, and its dipole orientation and emission energy must match what the circuit expects. Randomly dispersing emitters and hoping one lands in the correct spot is acceptable for a laboratory proof-of-concept. It is not a recipe for a technology platform. The same manufacturing instincts that demanded uniformity in printed display films reappear here: yield, alignment, repeatability. A quantum photonic chip cannot be a collection of lucky accidents.

The pressure of quantum information thus pulls the nanocrystal story into a different emotional register. Color, which once served as a friendly visible proxy for quantum confinement, becomes almost irrelevant. A single-photon dot might emit in the visible or in the near-infrared, depending on the application. What matters is the photon's character: its timing, its coherence, its statistical purity. The dot is judged not by how vivid it is but by whether it behaves like a disciplined quantum system rather than a noisy classical lamp.

And yet, the continuity with the earlier story remains in the most practical place: the surface. Surface chemistry is still the bottleneck, just with a sharper definition of what "good" means. A display can tolerate a small fraction of blinking emitters because millions contribute to the image. A quantum circuit often cannot tolerate even a small amount of intermittency, because the whole protocol may rely on every photon arriving when expected. The same blinking that was an annoyance in imaging becomes a show-stopper. The same Auger recombination that was a loss mechanism in LEDs becomes a source of uncertainty in computation.

One can feel the field's ambition in how it talks about nanophotonic coupling. Cavities and antennas are presented as a way to turn a single nanocrystal into a true component: a device that plugs into an optical network with defined input and output channels. That is the quantum version of industrial integration. Instead of printing inks and sealing films, one shapes electromagnetic environments and routes photons through waveguides. The goal is the same: to take a tiny, temperamental object and surround it with a structure that makes its behavior dependable.

A single photon is a remarkably fragile messenger. It can be absorbed by a speck of dust, scattered by a rough surface, or rendered useless for interference by a slight shift in frequency. Asking a colloidal emitter to supply such photons on command is a stringent test of how well nanomaterials can be tamed. When the test is passed, it is not only a victory of quantum mechanics; it is a victory of control—over surfaces, over noise, over the environment, over the narrow set of conditions in which one quantum event can be repeated with industrial regularity.

11.4 A Future Written in Sizes

At first glance, "color by size" sounds like a charming trick: make a crystal smaller, it glows bluer; make it larger, it glows redder. The deeper implication is more unsettling and more powerful. Geometry becomes a control parameter on the same level as composition. Instead of asking only *what* a material is made of, one can ask *how big* its active building blocks are, how sharply those sizes are distributed, and how precisely they are placed inside a larger structure. That is a new kind of materials language: a way of writing functions into matter without changing the periodic table.

That language, however, does not spread through society as a set of equations. It spreads through factories, standards, and rules. The future of nanocrystals depends not only on what can be made, but on what can be specified, measured, regulated, and trusted.

Consider something as ordinary as the word "red." In everyday life, it is a feeling. In manufacturing, it has to become a number. The moment a technology enters a supply chain, color must be communicated across continents and across contracts: a materials supplier says what they are selling, a device maker says what they need, and a retailer promises what a customer will see. That shared language is provided by standardized colorimetry frameworks developed by the Commission Internationale de l'Éclairage (CIE). These frameworks allow a spectrum to be reduced to agreed-upon coordinates that relate to human vision, and they underpin claims about "color gamut" and emission purity in displays.

Colorimetry is an infrastructure story with a surprisingly human beginning. In the early twentieth century, as electric lighting and mass printing expanded, it became obvious that subjective descriptions were not enough. Industries needed a way to guarantee that a batch of paint matched yesterday's batch, and that a lamp tested in one laboratory could be compared to a lamp tested in another. The CIE's 1931 system was a landmark not because it discovered new physics, but because it created a practical bridge between spectra and perception. It turned arguments about color into measurements that could be audited. Only with that bridge does "tunable emission" become a commercial property rather than an aesthetic anecdote.

Once nanocrystals become part of display products, they inherit this world of standardized claims. A material that emits at a particular peak wavelength is not automatically "the right green" in a device context. The perceived hue depends on the full emission profile, on how that emission interacts with filters, and on the device's white point. Manufacturers therefore talk in terms of chromaticity and gamut coverage, and they do it using the same CIE language across brands and borders. In that sense, standards do not merely describe technology; they *enable* it by making performance legible to everyone who touches the product.

The need for shared language does not stop at color. As nanocrystals push into new roles, the concepts themselves proliferate: quantum efficiency, coherence, antibunching, defect densities, photostability, spectral diffusion. Even the word "quantum" is used loosely in marketing and precisely in laboratories, and that mismatch can become costly when regulators, engineers, and investors interpret the same term differently. ISO vocabulary standards for quantum phenomena have been developed with a simple goal: facilitate communication across industry, regulators, and academia. Vocabulary standards rarely make headlines, yet they prevent a thousand small misunderstandings from hardening into product failures, legal disputes, or misplaced public fears.

This is a recurring pattern in technological history. The steam engine did not reshape society because a clever machine was built once. It reshaped society because pressure, safety valves, boiler plates, and inspection regimes became standardized enough that steam could be treated as dependable power rather than controlled danger. Electricity followed a similar path: voltage, frequency, plugs, and wiring codes turned a dazzling phenomenon into household infrastructure. Nanocrystal technologies are undergoing their own version of that transition. The "quantum" part is spectacular, but the future hinges on the unglamorous part: specifications, metrology, and repeatable manufacturing.

Manufacturing imposes its own definition of the frontier. In the laboratory, a breakthrough is often a record performance: brighter emission, narrower linewidth, a new composition. On a production line, the decisive metrics can be reproducibility, yield, and cost. Reviews of scale-up emphasize precisely this: success is defined by whether nanoscale physics can be translated into macroscale process control. A dot whose emission can be tuned by adjusting synthesis conditions is impressive; a dot whose emission can be tuned *predictably* across thousands of litres of production is transformative.

The "size knob" therefore migrates. It moves from a chemist's timing and intuition to a set of controlled variables in reactors and purification trains. It moves from "What is the average?" to "How stable is the distribution?" It moves from "Can it be made?" to "Can it be made, certified, shipped, printed, sealed, and warrantied?" Those questions can feel like constraints, but they are also the mechanism by which a clever material becomes an everyday component.

Displays provide a vivid example of how the future can be shaped by infrastructure rather than by a new core composition. Device roadmaps that combine μLED arrays with nanocrystal color conversion hint that major progress may come as much from integration, packaging, and manufacturing tooling as from a brand-new emitter. A blue μLED backplane is a formidable piece of engineering on its own; adding patterned conversion layers introduces requirements on film thickness, uniformity, and encapsulation that resemble the demands of microfabrication and printing. The bottleneck is often no longer "Can the nanocrystal emit?" but "Can it survive process steps and then hold performance under intense blue excitation for years?" That is not a question answered by a spectrum in a glovebox. It is answered by barrier layers, edge-sealing strategies, and quality control protocols that detect small drifts before they become large failures.

Regulation adds another layer of reality to the future written in sizes. Some of the most successful nanocrystal materials historically relied on elements that are politically and environmentally sensitive. The European Union's Restriction of Hazardous Substances (RoHS) framework shapes material choices and supply chains by restricting certain substances while allowing, through formal exemption procedures, temporary permission for specific applications. Those exemptions involve renewals and consultations; they are not a blank cheque. The result is that material selection is influenced not only by performance but by compliance trajectories.

This policy-to-plant linkage can be felt in design decisions that might otherwise be interpreted as purely scientific. A manufacturer may invest heavily in cadmium-free emitters not simply because they are intrinsically superior today, but because the regulatory and market landscape makes them a safer long-term bet. Conversely, a company might justify continued use of restricted substances in narrowly defined applications where performance is hard to match and where containment can be demonstrated. Either way, regulation is not merely an external constraint; it becomes an engineering input. It can accelerate the adoption of alternative compositions, encourage better encapsulation and containment strategies, and alter how companies think about end-of-life disposal and recycling.

Standards and regulation also change the public meaning of "safe." A nanocrystal embedded in a sealed display film is not the same risk as a soluble compound handled in a laboratory, yet public concerns often treat them as morally equivalent. That is where transparent measurement and clear vocabulary matter. If a device's safety relies on robust encapsulation and low ingress rates, then barrier performance becomes part of the safety story, not an internal manufacturing detail. Trust is built through demonstrable containment and consistent compliance, not through reassurances.

The most exciting extension of the size-to-function paradigm may be the one that forces an even stricter definition of "precision." Quantum-information directions suggest a frontier where controlling size and structure must deliver not only tunable emission but coherence. That is a profound shift in what it means to engineer a nanocrystal. In a display, a small spectral drift can be compensated by calibration, or it can be averaged across many emitters. In a quantum photonics device, the photon's frequency stability and indistinguishability can be central to whether an interference experiment works at all. Precision becomes less about hitting a target color and more about suppressing noise sources: spectral diffusion, charge fluctuations, and surface-related intermittency.

The analogy to the display world is instructive. Barrier layers and encapsulation determine whether a bright film remains bright. In quantum photonics, optical cavities and waveguides determine whether a single emitter becomes a useful source, and surface engineering determines whether the photons it emits are clean enough to interfere. Both worlds reward the same habit of mind: treat the surrounding architecture as part of the emitter, not as mere packaging. Both worlds demand repeatable manufacturing, not artisanal triumphs.

This expanded idea of precision can feel like a narrowing of freedom—more constraints, more tests, fewer materials that "qualify." But it also suggests why the underlying principle remains fertile. If size and shape can tune electronic structure, then they can tune far more than hue: absorption, charge transfer, radiative lifetimes, and interaction with photonic environments. The promise is not that every application will be won by the same family of nanocrystals. The promise is that geometry provides a new axis of design that can be adapted, with disciplined engineering, across many domains.

There is a subtle cultural shift embedded here. In older materials industries, a "material" was something you could hold: copper wire, silicon wafers, glass. In nanocrystal technology, the material often arrives as a liquid dispersion, becomes a printed film, and only then becomes part of a device. Its identity is distributed across synthesis, purification, ink formulation, deposition, and encapsulation. That makes the future of nanocrystals inseparable from the future of manufacturing methods that can handle such distributed identities: reactors that control particle populations, metrology that detects small defect-related changes, patterning techniques that maintain optical quality, and barrier technologies that turn sensitivity into durability.

The payoff, if it is achieved, is that "size" becomes a reliable engineering variable rather than a laboratory curiosity. One can then imagine design conversations that sound strangely straightforward: choose a nanocrystal size to set an emission energy; choose a surface treatment to set stability and interface compatibility; choose a photonic environment to set extraction efficiency or coherence; choose a barrier architecture to set lifetime. The complexity does not disappear, but it becomes organized, and organized complexity is what engineering is for.

A future written in sizes is therefore not a prediction of one dominant product. It is the claim that geometry has joined composition as a practical grammar for building devices. That grammar will be enforced by standards that make claims comparable, by regulations that shape material choices, and by manufacturing disciplines that treat variability as the central adversary. When those forces align, a nanocrystal stops being an object admired under a microscope and becomes a unit of capability: a small, repeatable piece of matter that can be trusted to behave the same way inside millions of devices, or inside a circuit where even a single photon counts.

www.ingramcontent.com/pod-product-compliance
Lightning Source LLC
LaVergne TN
LVHW010712061126
851857LV00014B/1145
* 9 7 9 8 2 5 6 2 8 9 2 9 4 *